Practical Information Security

Izzat Alsmadi • Robert Burdwell
Ahmed Aleroud • Abdallah Wahbeh
Mahmood Al-Qudah • Ahmad Al-Omari

Practical Information Security

A Competency-Based Education Course

 Springer

Izzat Alsmadi
Texas A&M University San Antonio
One University Way
San Antonio, TX, USA

Ahmed Aleroud
Department of Computer
Information Systems
Yarmouk University
Irbid, Jordan

Mahmood Al-Qudah
Yarmouk University
Irbid, Jordan

Robert Burdwell
Texas A&M University San Antonio
One University Way
San Antonio, TX, USA

Abdallah Wahbeh
Slippery Rock University of Pennsylvania
Slippery Rock, PA, USA

Ahmad Al-Omari
Schreiner University
Kerrville, TX, USA

ISBN 978-3-319-89143-9   ISBN 978-3-319-72119-4   (eBook)
https://doi.org/10.1007/978-3-319-72119-4

Printed on acid-free paper

This Springer imprint is published by Springer Nature
The registered company is Springer International Publishing AG
The registered company address is: Gewerbestrasse 11, 6330 Cham, Switzerland

Contents

The original version of this book was revised. An erratum to this book can be found at
https://doi.org/10.1007/978-3-319-72119-4_14

Chapter 1
Introduction to Information Security

Overview

Nowadays, security is becoming number one priority for governments, organization, companies, and individuals. Security is all about protecting critical and valuable assets. Protecting valuable and critical assets, whether they are tangible or intangible, is a process that can be ranged from being unsophisticated to being very sophisticated. Security is a broad term that serves as an umbrella for many topics including but not limited to computer security, internet security, communication security, network security, application security, data security, and information security. In this chapter, and following the scope of the textbook, we will discuss about information security and provide an overview about general information security concepts, recent evolutions, and current challenges in the field of information security.

Knowledge Sections

In the knowledge section, we will cover concepts and terminologies related to information security. More specifically, we will discuss about information taxonomy, security goals, security countermeasures, and security development lifecycle. In addition, we will discuss about recent evolutions in the field of information security as well as current challenges faced by governments, organizations, companies, and individuals.

Information security, sometimes referred to as InfoSec, is defined as processes, methodologies, standards, mechanisms, and tools which are designed and implemented for the purposes of protecting information from unauthorized access, use, modification or destruction, in order to ensure confidentiality, integrity, and availability of information.

© Springer International Publishing AG 2018
I. Alsmadi et al., *Practical Information Security*,
https://doi.org/10.1007/978-3-319-72119-4_1

Information security in this context is a general term that can apply to any format of information whether it is electronic or physical, whether information in transit, in rest, or in motion, whether information is created, stored, destructed, processed, or transmitted, or whether information is classified as sensitive information, restricted information, or unrestricted (public) information.

In order to facilitate reading through the chapters of this book, it is good idea to understand some basic terminologies and concepts related to information security. In the subsequent sections, we will discuss about concepts related to information taxonomy, security goals, security counter measures, and security development lifecycle.

Information Taxonomy

In order to understand information security, it is a good idea to learn about information taxonomy. Information taxonomy is all about describing and classifying information based on a number of criteria including but not limited to format, state, location, sensitivity, etc.

Information can take many different formats. Some information is available in a physical format such as textbooks, journals, newspapers, policies, and contracts. Other information is available in an electronic format that includes text, audio, and videos such as Webpages. Understanding the format of information you are working with is crucial to achieve the objectives of information security.

Information State

Another important aspect of information is the state of information. In order to achieve the objectives of information security and identify the appropriate counter-measures for such information, it is crucial to know about different states of information. Such states include information creation, information processing, information storage, information transition, and information destruction. At any time, information may be in one of the aforementioned state. Also, the state of information might change from one to another during its lifecycle.

- Information creation: Information creation is the process of generating content that convey a message and shared via different methods. The information creation process is an iterative and incremental process. In order to achieve information security goals, information need to be protected while it is being created using different applications.
  During the information creation process, the integrity of the information must be ensured by checking the completeness and correctness of such information.

In addition, we need to ensure the provenance of information via audit-ability and accountability. Finally, we need to decide and agree on the correct level of information sensitivity to maintain confidentiality and privacy (Cherdantseva and Hilton 2013).

- Information storage: Once data is created, it need to be stored in order to make accessible to other applications and processes. Information is stored onsite using storage devices such as hard drive, flash drives, or network storage or offsite such as another physical location or using the cloud. Once information is created and stored, it must be protected from unauthorized access, whether such information is located onsite or offsite, using different security measures such as encryption, and passwords. Other kinds of information in physical formats must be protected using physical security measures such as magnetic stripe ID cards and locks.

- Information processing: Information processing is the process transforming information from one format to another. In some cases, processing information can cause information losses. Such losses can be intentional or intentional. Information need to be protected it is processed from one format to another. To do so, organization need to have the appropriate security measures in place in addition to a good incident response plan in case an information processing error occurs.

- Information transition: Information transition is the process of sharing information or giving others access to information. Such information is exchanged using a number of open and proprietary protocols (Deghedi 2014). Information need to be protected while it is in transition over public or private networks. One widely used technique to protect information during transition is encryption.

- Information destruction: After a period of time, information available on different storages and formats are outdated and outlived their purposes, as a result, such information as well as the place where is stored need to be disposed and destroyed. This process is called is information destruction. One of the important goals of information security has to do with protecting information when organizations decide to get rid of it. The destruction process need to be controlled, audited and executed in a lawful way (Cherdantseva and Hilton 2013). The selected method of destruction is highly dependable on the type and sensitivity of information.

Information Location

Risk to information as well as the required security measures depends on the current location of information (Cherdantseva and Hilton 2013). For example, information located on laptop computers may requires different security measures than information located on a desktop computer located in the organization's office. Also, information that resides within the organization and processed internally requires different security measures than information in motion.

- Information in motion refers to information that is actively moving from one location to another. Such information is vulnerable to security risks more than information at rest since such information is being transmitted over public or private networks. In some cases, such information is moving back and forth between local storages and the cloud. In both cases, information protection for information in motion is crucial to achieve information security goals. One approach to protect information in motion is to control access to such information. Another mechanism is use encryption and making sure that all connection over which the informing is transmitted is encrypted.
- Information at rest refers to information that is not moving from one location to another. Such information is stored on hard drives, flash drives, or being archived. Protecting information at rest is easier than information in motion since you only need to worry about information within your organization, or information on your personal devices. In most cases, data at rest are more valuable than data in motion, which causes such information to be one of the targets for attackers. Protecting information at rest can be achieved by following a number of security measures such as firewalls and anti-virus programs. In some cases, encrypting the information located at your devices is considered the most effective security measure to protect information at rest.

Information Sensitivity

Information must be protected based on its value as well as the likelihood that such information may be targeted for unauthorized disclosure. In general, information can fall in one of three categories based on its sensitivity: confidential information, private information, public information. This classification of information sensitivity is independent of the format and status. The only difference between these categories is the likelihood, duration, and the level of harm incurred in case an unauthorized access occurs.

- **Confidential information:** Confidential information represents information that ca results in significant level of risk when unauthorized disclosure, alteration or destruction of that information occurs. Examples of confidential information include electronic medical records, financial information, and credit card transaction. In most cases unauthorized access to such information can results in a significant monetary loss for the owner of the information as well as long-term harm. As a result, confidential information should be maintained in a way that allows only authorized people to access it. In such context access controls can be implemented in a way that allows access to data based on roles or on need basis. Keep in mind that the highest level of security controls need to be applied to confidential information. Restricted information is one of the sensitive categories of confidential data. Restricted information is defined as "information that cannot be disclosed to an unauthorized organization or to one or more individuals" (Sengupta 2011).

- **Private Information:** Private information represents information that can result in a moderate, minor risk in case unauthorized disclosure occurs. In general, all information that is not classified as confidential information or public information is considered private information. A reasonable level of security controls should be applied to private information (Ivancic et al. 2015).
- **Public information:** Public information represents information that can result in a little or no risk in case unauthorized disclosure occurs. In most cases, public information is available to anyone who needs access to it. Examples of public information include but not limited to press release, maps, directories, and research publications.

Security Goals

The CIA-triad - confidentiality, integrity, and availability – is no longer considered as an adequate and complete set of security goals. The CIA-triad does not cover new threats that emerge in the collaborative de-parameterized environment (Cherdantseva and Hilton 2013). A new list of security goals is discussed in this section that accounts for all current information security threats.

- **Confidentiality:** in the context of information security, confidentiality means that information should be secured and only authorized users may access and read such information. Confidentiality is very important to individuals as well as organization since violating this goal can lead to devastating consequences. Security threats to information confidentiality are malware, social engineering, network breaches. Since confidentiality is really important, mechanism for protecting information should be in place. Security measures for maintaining information confidentiality include but not limited to cryptographic techniques and access controls.
- **Integrity:** integrity is all about making sure that the information has not been modified or corrupted. Data integrity covers data in storage, during processing, and while in transit (Stoneburner et al. 2001). Information integrity involves ensuring information non-repudiation and authenticity. In this context, source integrity plays an integral role in validating information integrity. Source integrity is defined as making sure that the sender of the information is who it is supposed to be (Matteucci 2008). Spoofing is one of known threats to integrity where the attacker deceives the receiver and supplies incorrect information. Information integrity protection mechanisms are grouped into preventative mechanisms and detective mechanisms. Example techniques for protecting information integrity include digital signatures and hash algorithms.
- **Availability:** availability is all about ensuring timely and reliable access to and use of information (Stoneburner et al. 2001). The integrity and conditionality of information is worthless if such information is not available for the intended users. Threats to information availability s not only technical one such as Denial

Fig. 1.1 Identification and
authentication using
username and password
(ITPRO 2014)

of Service (DoS) attack, they also include natural and manmade disasters such as
tornados and fire. One of the key security measures to protect against threats to
information availability is backups.

- **Identification**: Identification is the first step in the identify-authenticate-
authorize sequence that is carried daily whenever a user is accessing information.
The identification all about calming that you are somebody. For example, one
can claim that he is John Doe. In the context information security, identification
is similar to providing a username or something else that uniquely identifies a
user. Nowadays, the most commonly used identification method is user ID.
- **Authentication**: authentication is the second step in the identify-authenticate-
authorize sequence that verifies the authenticity of the claimed identity. During
authentication, the user proves that you are indeed the one you claim to be.
Authentication occurs often as a prerequisite to allowing access to resources in
an information system (NIST 2013). In case of John Doe, the ID card is used to
authenticate the claim he is John Doe not someone else. In the context of information security, there is different ways to authenticate users such as something
you know (passwords), something you have (smart card), or something you are
(biometrics). In certain situations, and depending on the sensitivity of the information, a combination of these authentication methods can be used (Fig. 1.1)
- **Authorization**: access privileges granted to a user, program, or process or the act
of granting those privileges (Dufel et al. 2014). Once the user is identified and
authorized, they are assigned a set of permission and privileges - known as
authorization – that defines what they can do with the information and the system. Unlike identification, which requires some sort of username, and authentication, which requires for example password, authorization is implemented part
of the security policy within the organization. The most commonly used approach
for granting permissions once users are authorized is role-based access control in
which permissions are associated with roles, and users are made members of
appropriate roles (Sandhu et al. 1996).
- **Accountability**: the security goal that generates the requirement for actions of
an entity to be traced uniquely to that entity. This supports non-repudiation,
deterrence, fault isolation, intrusion detection and prevention, and after-action
recovery and legal action (Stoneburner et al. 2001).
- **Privacy:** privacy is the process of restricting access to subscriber or relying party
information in accordance with federal law and agency policy (Kuhn et al. 2001).

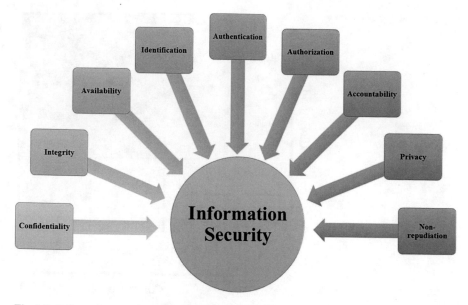

Fig. 1.2 Information security goals. Developed by author

In the context of information security, privacy is critical in case the information identifies human beings. Such information includes social security number, name, and address. Information security policies play a major role in protecting personal information, and how such information is collected and processed.

- **Non-repudiation:** is the security service by which the entities involved in a communication cannot deny having participated. Specifically, the sending entity cannot deny having sent a message, and the receiving entity cannot deny having received a message (NIST 1994) (Fig. 1.2).

Security Risks, Threats, and Vulnerabilities

In the field of information security, some of the terminologies or concepts used are misunderstood or in some cases are used synonymously. Some security concepts are closely related and worth being examined together. Widely related terms in information security include risk, threat, and vulnerability. In this section, we discuss about these concepts briefly.

- **Risk**: risk is the likelihood that assets are being threatened by a particular attack. System-related security risks are those risks that arise from the loss of confidentiality, integrity, or availability of information or systems and reflect the potential adverse impacts to organizational operations, organizational assets, individuals, other organizations, and the Nation (Michael et al. 2017) (Fig. 1.3)

Fig. 1.3 Risk assessment
(Lee 2014)

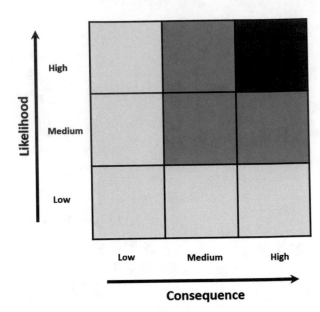

Table 1.1 Risk rating system

Risk	Risk rating
≥261	Very high
201–260	High
141–200	Medium high
101–140	Medium
61–100	Medium low
31–60	Low
1–30	Very low

When it comes to risks, risk assessment is a critical step that considers the potential threats, the impact it causes, and the probability it occurs. Based on the risk assessment, the appropriate security measures and policies are established and be in place. Table 1.1 shows the risk rating system to cinder when performing risk assessment.

- **Threat**: a threat refers to the sources, capabilities, or methods of an attack that attempts to exploits vulnerability in the target system, with the potential to adversely impact organizational operations, organizational assets, individuals, or other organizations via unauthorized access, destruction, disclosure, modification of information (Michael et al. 2017). Threats usually take the form of worms, viruses, and Trojan horses. To detect threats, threat assessment that focuses on assessing threats and choose the best countermeasures is usually done part of the organization wide security plan. Table 1.2 shows the threat rating scales.

Table 1.2 Threat rating scale

Very high	10
High	8–9
Medium high	7
Medium	5–6
Medium low	4
Low	2–3
Very low	1

Table 1.3 Vulnerability rating scale

Very high	10
High	8–9
Medium high	7
Medium	5–6
Medium low	4
Low	2–3
Very low	1

Table 1.4 Assets value

Very high	10
High	8–9
Medium high	7
Medium	5–6
Medium low	4
Low	2–3
Very low	1

- **Vulnerability**: vulnerability is flaw in the design or configuration of software that has security implications. Security breaches starts with an attacker scanning the system for potential vulnerability, and then use such vulnerability to carry a security attack. Testing for vulnerabilities is crucial for maintaining ongoing security. A variety of organizations maintain publicly accessible databases of vulnerabilities (Mell et al. 2005). Table 1.3 shows the vulnerability rating scale.

- **Asset:** An asset is a resource of value requiring protection. An asset can be tangible or intangible. Asset value is the degree of debilitating impact that would be caused by the incapacity or destruction of assets. Table 1.4 shows the assets values scale.

The measure of a risk can be determined as a product of threat, vulnerability and asset values (Vacca, 2012):

$$Risk = Threat^* \ Vulnerability^* \ Asset.$$

Security Countermeasures

A countermeasure is an action, device, procedure, or technique that counters a threat by eliminating or preventing it, minimizing the harm it cause, or discovering and reporting it so that corrective action can be taken (Ben Othmane et al. 2014). Nowadays, technical countermeasures are not capable to address information security threats alone. Other Information security countermeasures of different natures such as organizational, human, and legal should be in place along with the technical one (Cherdantseva and Hilton 2013).

- **Technical Countermeasures:** refer to technical means designed to achieve information security goals. For example, cryptography, biometrics, digital signature, firewall, intrusion detection system, antivirus, etc.
- **Organizational Countermeasures:** refer to administrative activities which aim to build and maintain a secure environment where selected security countermeasures may be effectively implemented and managed. For example, security strategy, security policy, audit, etc.
- **Human Countermeasures:** address the impact of the human-factor on information security. For example, education, awareness, training, certification, ethics, culture, motivation, etc.
- **Legal Countermeasures:** refer to the use of the legislation for the purposes of information protection. For example, service-level agreements, copyright law, etc.

Security Education, Training and Awareness

Security education, training and awareness are considered crucial to organization's security framework, and must be customized to the structure and activities of that framework. A security education, training and awareness program is an "educational program that is designed to reduce the number of security breaches that occur through a lack of employee security awareness" (Whitman and Mattord 2013).

The purpose of information security awareness, training, and education is to enhance security by raising awareness of the need to protect system resources; developing skills and knowledge so system users can perform their jobs more securely; and building in-depth knowledge as needed to design, implement, or operate security programs for organizations and systems. (Michael et al. 2017).

Information Security Challenges

Despite the fact that many technological and procedural solutions have been developed for information security and more are in progress, security issues are a great challenge to most organization.

Long time ago, information security was considered a pure technical problem, however, nowadays, security challenges are more related to management and individuals. Top management should be always aware of the importance of information security policy development and implementation. In addition, the organization structure poses an important challenge in managing information security. Technology is not enough to protect against security issues. Organizations are looking for more reliable protection mechanisms that consider technical, human, social, and organizational aspects, not only the technical one when implementing an information security plan. Such balanced approach to information security is referred to as socio-technical information security needs. Humans are the most critical element in information security management. According to (Soomro et al. 2016), causes of information and data breaches are related to lost paper files (38%), misplaced memory devices (27%), and hackers (11%). So, the problem of information security nowadays, is more socio-technical rather than being pure technical problems.

Another challenge related to information security is the development and implementation of an effective information security policy. Even in case such policy exists, it is necessary to make sure that awareness and training regarding such policy takes place, because an information security policy without training and education is useless. Finally, compliance with information security policy through training and education is challenging. In some cases, employees are aware of the existence of an information security policy, received the necessary training and education, but they do not comply with the policy terms. In such case, issues might arise and security might be compromised.

Another information security challenge is related to cloud computing. Cloud computing is a new emerging concept in computing technology that utilizes internet and remote servers to maintain data and applications. Security and privacy issues present a strong barrier for users to adapt into cloud computing. Issues related to information location by users pose a challenge to cloud computing security. Another issue is related to information segregation since information on the cloud is stored on devices that are shared with other cloud users. Finally, recovery is another important issue and the biggest challenge is the need for a contingency plan that cloud users can trust when information is threatened by security risks.

The strategic alignment of business and IT/IS for optimization of information security is another important challenge because information security is highly dependent on the strategic business IT alignment (Soomro et al. 2016). The lack of alignment will lead to the fact that security policies and strategies will not reflect business's needs. As information technologies are integrated with almost all businesses, it becomes more critical. As a result, the security of such technology and related assets should not be left out to technical people; management should be involved in the process.

The increased number of connected devices in Internet of Things (IoT) has been translated into security risks and poses new security challenges. Almost all security challenges are found in the IoT. As a result, all security goals need to be considered part of the security fundamentals. When it comes to IoT, we need to make sure that things are securely connected to the network, controlled, and only accessible by authorized users. In addition, connected devices generate information, where such information need too collected, analyzed, stored and presented in a secure manner. Finally, there is a need to make sure that things are not collecting private and personal information, and that information collected by such things are not leaking, used by unauthorized users miscellaneously (Elkhodr et al. 2016).

Skills Section

CIA Triad – Confidentiality, Integrity and Availability

1. In some situation, some of the information security goals are more important than others. Given the basic security goals, confidentiality, integrity and availability, give examples of situations or cases where one goal is highly important than others.

Information Security Cases

1. The following cases are based upon real situations (WikiBooks 2017).

   - An employee was using a flash drive to move the company employees' information. Information is about 9000 employees including but not limited to names, addresses, social security numbers, telephone numbers, job related information, and date of birth. This flash drive went missing. No evidence or report indicates that information has been compromised or used inappropriately.
   - A laptop containing sensitive information about the company employees went missing. The management believes the laptop has been stolen. All known about the laptop is that it is protected with a username and password. Also, information has been processed and stored in a format that would not be easily accessible. No evidence or report indicates that information has been compromised or used inappropriately.
   - A student installed a key-logger application on labs' computers in a university in order to gain access to students and faculty username and passwords. The student used the obtained usernames and passwords to access the learning management system to change students' grades.

   For each case study, answer the following questions:

   (a) What should be the very first course of action?
   (b) Should the public be informed about the situation? If so, how will their trust be regained?
   (c) What steps should be taken to prevent similar attacks in the future?

(d) What are the ethical issues of this situation?

(e) How should employee/student be dealt with if they were the people initiating the attack?

Applications Section

Information Security Basics: Testing Your Anti-virus Software Application

If you ever want to test your antivirus software, you can use the EICAR test file. The file is not an actual virus; it is simply a text file that contains a string of harmless code that prints the text. The file is mainly used to test your anti-virus application and make sure that it will detect the file.

1. You can download an EICAR test file from the EICAR website. Click on the Anti-Malware Test File Link (alternatively, go to http://www.eicar.org/85-0-Download.html).

2. Before downloading nay file type, make sure to read the instructions. Using the DOWNLOAD link on the left side of the page, you can download any of several forms of EICAR test file (Fig. 1.4)

3. Try downloading each form of the EICAR test file and note how your anti-virus software responds. If your anti-virus software application is active and working, it should display a pop-up a window when you try to download and save the eicar.com or eicar.com.txt file. An example virus detection pop-up window is shown in Fig. 1.5.

Information Security Basics: Vulnerability Scanning

Personal Software Inspector (PSI) is a security scanner which identifies programs that are insecure and need updates. It automates the updating of the majority of these programs, making it a lot easier to maintain a secure PC. It automatically

Download area using the standard protocol http

eicar.com	eicar.com.txt	eicar_com.zip	eicarcom2.zip
68 Bytes	68 Bytes	184 Bytes	308 Bytes

Download area using the secure, SSL enabled protocol https

eicar.com	eicar.com.txt	eicar_com.zip	eicarcom2.zip
68 Bytes	68 Bytes	184 Bytes	308 Bytes

Fig. 1.4 EICAR test files (EICAR, http://www.eicar.org/85-0-Download.html)

Fig. 1.5 Virus detection
pop-up window

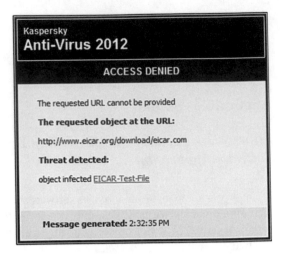

detects insecure programs, downloads the required patches, and installs them accordingly without further user interaction.

In this project, you will perform vulnerability scanning using the Secunia PSI tool and identify which patches are missing. Download Secunia PSI at http://download.cnet.com/Secunia-Personal-Software-Inspector/3000-2162_4-10717855.html. If the resource is unavailable, use Google to locate the Secunia PSI software. Install PSI on your system and begin a scan.

Answer the following questions:

1. Allow the scan to complete. Provide a screen shot showing your scan was successful
2. What was your Secunia System Score?
3. Which patches were identified as missing or what programs were listed as insecure?

Questions
- Why information security is very important.
- Describe the levels of information sensitivity and provide two examples about each level.
- Explain the concept of information destruction. Give one real life scenario where information destruction is considered critical.
- What is the difference between information in motion and information at rest?
- For each type of security goals, give two examples about security measures to achieve each goal.
- Explain how confidentiality, integrity, and availability complement each other.
- Explain the identify-authenticate-authorize process. Give an example to demonstrate the process.
- Explain the relationship between risks, threats, and vulnerabilities.
- Briefly explain the concept of information non-repudiation.

- What is the difference between technical, organizational, human, and legal countermeasures? Give one example about each countermeasure category.
- Explain the concept of security education, training and awareness program.
- Briefly explain the information security challenges.

References

Ben Othmane, L., Angin, P., Weffers, H., & Bhargava, B. (2014). Extending the agile development process to develop acceptably secure software. *IEEE Transactions on Dependable and Secure Computing, 11*(6), 497–509.

Cherdantseva, Y., & Hilton, J. (2013). *A reference model of information assurance & security.* Paper presented at the Availability, reliability and security (ares), 2013 eighth international conference on.

Deghedi, G. A. (2014). *Information sharing as a collaboration mechanism in supply chains.* Paper presented at the Information and Knowledge Management.

Dufel, M., Subramanium, V., & Chowdhury, M. (2014). Delivery of authentication information to a RESTful service using token validation scheme: Google Patents.

EICAR (n. d.). EICAR test files, Retrieved from http://www.eicar.org/85-0-Download.html

Elkhodr, M., Shahrestani, S., & Cheung, H. (2016). The internet of things: New interoperability, management and security challenges. *arXiv preprint arXiv:1604.04824.*

ITPRO. (2014). Russian cyber gang steal 1.2 billion sernames & passwords, Retrieved from http://www.itpro.co.uk/security/22838/russian-cyber-gang-steal-12-billion-usernames-passwords

Ivancic, W. D., Vaden, K. R., Jones, R. E., & Roberts, A. M. (2015). Operational concepts for a generic space exploration communication network architecture. NASA, online, https://ntrs.nasa.gov/archive/nasa/casi.ntrs.nasa.gov/20160013860.pdf

Kuhn, D. R., Hu, V. C., Polk, W. T., & Chang, S. J. (2001). Introduction to public key technology and the federal PKI infrastructure. Retrieved from. NIST, https://csrc.nist.gov/publications/detail/sp/800-32/final

Lee, J. (2014). An enhanced risk formula for software security vulnerabilities. *ISACA Journal, 4.*

Matteucci, I. (2008). *Synthesis of secure systems.* PhD thesis, University of Siena. https://www.semanticscholar.org

Mell, P., Bergeron, T., & Henning, D. (2005). Creating a patch and vulnerability management program. *NIST Special Publication, 800,* 40.

Michael, N., Kelley, D., & Victoria, Y. P. (2017). *An Introduction to Information Security* (pp. 800–812). NIST online, http://nvlpubs.nist.gov/nistpubs/SpecialPublications/NIST.SP.800-12r1.pdf

NIST. (1994). *Federal Information Processing Standard (FIPS) 191.* National Institute of Standards and Technology (NIST). https://csrc.nist.gov/csrc/media/publications/fips/140/2/final/documents/fips1402.pdf

NIST. (2013). *Security and privacy controls for federal information systems and organizations.* National Institute of Standards and Technology (NIST). https://csrc.nist.gov/csrc/media/publications/sp/800-53/rev-4/archive/2013-04-30/documents/sp800-53-rev4-ipd.pdf

Russian cyber gang steal 1.2 billion usernames & passwords, Retrieved from http://www.itpro.co.uk/security/22838/russian-cyber-gang-steal-12-billion-usernames-passwords

Jaewon Lee, An Enhanced Risk Formula for Software Security Vulnerabilities. ISACA Journal Volume 4, 2014.

Sandhu, R. S., Coyne, E. J., Feinstein, H. L., & Youman, C. E. (1996). Role-based access control models. *Computer, 29*(2), 38–47.

Sengupta, A. (2011). Method for processing documents containing restricted information: Google Patents.

Soomro, Z. A., Shah, M. H., & Ahmed, J. (2016). Information security management needs more holistic approach: A literature review. *International Journal of Information Management, 36*(2), 215–225.
Stoneburner, G., Hayden, C., & Feringa, A. (2001). *Engineering principles for information technology security (a baseline for achieving* security). Retrieved from NIST, https://csrc.nist.gov/publications/detail/sp/800-32/final
Vacca, J. R. (2012). *Computer and information security handbook* (2nd ed.). Cambridge, MA: Newnes\Morgan Kaufmann.
Whitman, M., & Mattord, H. (2013). *Management of information security* (4 ed.). Nelson Education\Cengage Learning.
WikiBooks. (2017). Information security in education - case studies. Retrieved from https://en.wikibooks.org/w/index.php?title=Information_Security_in_Education/Case_Studies&stable=1

Chapter 2
The Ontology of Malwares

Overview

Absolute security is almost impossible. On a daily basis the security of many systems is compromised. Attackers utilize different techniques to threaten systems' security. Among different threats to systems' security, malware poses the highest risk as well as the highest negative impact. Malware can cause financial losses as well as other hidden cost. For example, if a company system has been compromised, the company could suffer negatively on the reputation and trust level from a publicized malware incident. This chapter provides a detailed description about different malware categories and how to protect against each type.

Knowledge Sections

In the knowledge section, we will cover concepts and terminologies related to different types of malware. The chapter provides a taxonomy of different malware including adware, spyware, viruses, worms, Trojans, Rootkits, Backdoors, keyloggers, rogue security software, ransom ware, browser hijackers, and logic bombs. The chapter covers detailed information about each category as well as guidelines for protecting against each malware type. Table 2.1 summarizes the major malware categories.

© Springer International Publishing AG 2018
I. Alsmadi et al., *Practical Information Security*,
https://doi.org/10.1007/978-3-319-72119-4_2

Table 2.1 Malware classes (Grégio et al. 2012)

Class	Behavior
Evader	Removal of evidence Removal of registries AV engine termination Firewall termination Notification of updates termination Language checking
Disrupter	Scanning of known-vulnerable service E-mail sending (spam) IRC/IM known port connection IRC/IM unencrypted commands
Modifier	Creation of new binary Modification of existing system binary Creation of unusual Mutex Modification of the name resolution file Modification of the browser proxy settings Modification of the browser behavior Persistence Download of known malware Download of unknown file Driver loading
Stealer	Stealing of system/user data Stealing of credentials or financial data System/user information reading Process hijacking

Spyware

Spyware is category of computer programs that tracks and reports computer users' behavior and internet habits without users' consent for marketing or illegal purposes (Wang et al. 2004). In general, Spyware is not designed to harm your computer. However, Spyware is not a simple tracking software. Spyware tracks your search habits, collects personal information, install software, redirect your Web browser, change your device settings, slow down your device, and pop-up ads. Nowadays, most Spyware targets and affect Windows operating system.

In general, Spyware are usually hidden and are difficult to detect. In some cases, Spyware is part of genuine software, while others come from malicious sources. Spyware infect devices by deceiving the users, for example clicking on a link or a button on a pop-up window; or by exploiting some vulnerabilities in the system, for example installing a piece of software from un-trusted sources.

Adware is considered a type of Spyware which are considered software that displays advertisements that bothers users and make them unpleased while they are surfing the Web. In case adware are malicious, they can exhibit the behavior of viruses, worms, or spyware. A tracking cookie, a subtype category under adware which represent a data structure used to store information about user's sessions on the Web, which can be used to track user's behavior on the Web. Tracking cookies are mainly used to track behaviors and link them to the user's personal data, which in turn is used to collect data and then sell it to marketers or cybercriminals (Goertzel 2009).

The followings are considered ways in which Spyware affect devices:

- **Piggybacked software installation**: In some cases, installing an application will end up with installing a Spyware on your device. Spyware installation will be part of the regular installation procedure of the downloaded software. In most cases, this happen part of installing free software application that users find as alterative to the paid one. The installed spyware will monitor and even modify your data and information (Hasan and Prajapati 2009).
- **Drive-by download**: Most users have experienced pop-up Windows while they are browsing the Web. In most cases, these pop-ups try to download and install Spyware on your device. Drive-by download usually occurs by exploiting the user's browser and download and install the Spyware without users' consent or knowledge. In other cases, the only message you will receive is a standard message with the name of the software and it is okay to install it (Sood and Enbody 2011) (Fig. 2.1).
- **Browser add-ons**: Add-ons are software that extends the functionality of your browser. Most browsers such as Chrome, Firefox, Opera, Safari support add-on – sometimes called plug-ins. Add-ons come in a toolbar, search box, or an icon format. In most cases, they do enhance the browser functionality and perform the work they are intended to do, at the same time they might include Spyware.
- **Masquerading as anti-spyware**: many users got deceived by software application that works as a toolbar or even as a software for detecting and removing spyware. (Gralla 2005).
- **ActiveX control**: This involves downloading and installing an active Control whenever it is required by a Website. In some cases, this method is not used for displaying dynamic content, but to deceive the user into downloading an installing a Spyware rather than ActiveX control (Chien 2005).

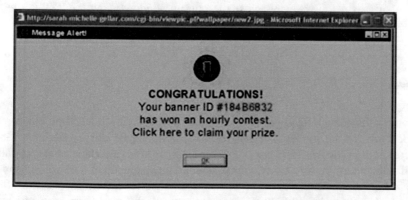

Fig. 2.1 Pop-up window tries to install spyware (http://sarah-michelle-gellar.org/)

Spyware can do many things once they are installed on your device. Spyware can run software application in the background that can slow down your devices by consuming the memory and processing power. Spyware can also generate engine, and even modify the too many pop-up ads that can slow down your browser, even make it useless. In some cases, it can show up an ad every time you start your browsers. Some Spyware can redirect your browser web-page, redirect your search results, change your default search engine, change Internet settings, and even modify the dynamically linked libraries used to connect to the Web. The worst type of Spyware is the one that records keyboard usage, access browsing history, and password saved on your browser.

The followings are considered ways to protect you against Spyware:

- **Use a spyware scanner or anti-spyware**: Many software products are available on the Web for detecting and removing Spyware from your device. Some of these applications are free while others are subscription based. They are effective as an anti-virus application and can provide real time detection and protection. The only thing you need to care about is the trustworthy of the source from which you are downloading the application.
- **Use a pop-up blocker**: Most browsers have the ability to block ads and pop-ups while you are browsing the Web. This function can be running all the time or can be conFigured to alert the user whenever a pop-up is required or the Website needs to be opened in a different Window. Also, you can setup a list of trusted Website from which pop-ups are allowed.
- **Disable Active-X**: Since some Spyware take advantages of your Active-X setting to trick you into installing malicious code instead of Active-X, then it is a good idea to disable Active-X on your browser.
- **Check software before installation**: Be careful when a Website asks to install an add-on or asks you to download and install a software application, especially if it is not something you recognize such as flash player and java engine.
- **Close pop-up window using "X" icon**: Whenever you receive a message in the form of pup-up window that has a number of options like "Yes" No", "Install" "Cancel", then make sure that you do not click on either options, especially when you are suspicious. Always close the pop-up window using the "X" at the top right corner.

Adware

Adware software applications are designed for displaying advertisement for the purpose of making money. Advertisements are displayed during the installation process of a particular software or in some cases in the interface of the installed application itself. In most cases, advertisement are displayed in the form of pop-up ads, and in other cases are based upon information gathered from users actions from devices (Levow and Drako 2005). Adware can risk the disclosure of confidential

Fig. 2.2 Infinite pop-ups

data by unknowingly redirecting individuals to "lookalike" Web sites (Gordon 2005). Not all adware is horrible, however, some of them can alter your security settings and track your activities and display advertisements. The problem in this case, is that such breaches by adware can be exploited by other kinds of threats.

Adware acts on devices in different ways (Lemonnier 2015):

- **Infinite Pop-ups**: A stream of endless pop-ups, while other times open one while closing others in an infinite loop (Fig. 2.2)
- **Spying**: Adware can track activity online with the purpose of targeting you with relevant advertisements.
- **Man-in-the-middle attacks**: Adware can act like man-in-the-middle by redirecting your traffic through their system to display advertisements. This even happen over what supposed to be secure connections.
- **Slowing down your device**: Adware displaying advertisements can slow down your devices' processing power and consume your memory and slow down device's performance.
- **Eating up your data**: Adware can consume you data plan, in case you have limited plan, same as listening to music or watching video, or eve browsing the Web.

Adware can infects your device in different ways (Berberick 2016):

- **Freeware or shareware**: Installing a freeware or shareware software can end up with Adware installed part of the process. This is considered a legitimate way to generate advertisements and make profit with the objective of funding the software.
- **Infected websites**: Visiting a Website can end up with installing Adware on your device. This occurs by exploiting a vulnerability via browser and installing the Adware.

Users can protect themselves from Adware and remove them in different ways:

- Blocking scripts from running on your browser can help protecting your device from Adware. However, this can also end up blocking other useful scripts since Adware are written in the same scripting language like other legitimate software
- Adware removal software is another way to protect against Adware since some types of adware are written in ways similar to viruses. In most cases, anti-virus suites protect against adware.
- Make sure that your browser is up to date and all security patches are installed. Also, make sure that your firewall is on when using the Internet.

Rootkits

A Rootkit is a malicious concealed software application that is designed to provide unauthorized access to devices or software application. Rootkit is composed of root and kit. Root is the admin account on Unix-like operating systems, and kit refers to the implementation of the tool. Rootkits are associated with Trojans, viruses, and worms that conceal their actions from users and systems' processes (Schmidt et al. 2008).

Rootkits can be installed on the victim device automatically or an attacker can install it once they have access to the victim device via the administrator account. Once Rootkit is installed it can become hidden and maintain the privilege access as administrator (Medley 2007). This means that the Rootkit has complete access to the software installed on the victim device as well as other setting making things easy to modify without detection.

Similar to Adware, Rootkits can be installed part of a legitimate process or malicious code. Rootkits can be installed while installing a legitimate software application. Some technologies use the Rootkit in order to work on devices. For example, Daemon Tool, a free imaging software application uses the following Rootkit detected by anti-virus software application

C:\Windows\System32\drivers\al887uj6.sys

Fig. 2.3 Computer
security rings (https://
www.pinterest.com/pin/
194288171397349001/)

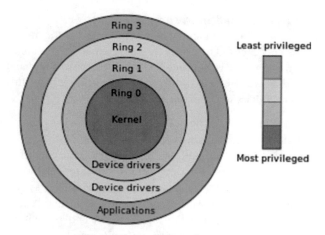

Even after the removal of the Rootkit, the same file will be restored again by the application in order to work. Rootkits can be also installed on devices maliciously through suspicious email attachment, plug-in, or even when installing other applications (Fig. 2.3).

A Rootkit falls into one of five major categories:

- **User mode:** User-mode Rootkits run (Ring 3) along with other applications rather than low-level system processes. User-mode Rootkits have different methods and approaches for modifying the behavior of existing application interfaces. (Kapoor and Sallam 2007).
- **Kernel mode:** Kernel-mode Rootkits run (Ring 0) with the highest operating system privileges. This is done by replacing, modifying, or adding codes to part of the operating system, including both the kernel and associated device drivers (Muttik 2014). This category of Rootkits has unrestricted security access and difficult to write. If kernel-mode Rootkits written with bugs in the code, ten it can be easily discovered by countermeasures.
- **User-mode/kernel-mode hybrid Rootkit**: In some cases, Rootkit developers get the most out of the developed Rootkit when they combine the characteristics of both user-mode and kernel-mode Rootkits. Nowadays, this is the most popular and effective type of Rootkit.
- **Bootkits:** A bootkit is a variant of kernel-mode Rootkit that can infect startup code like a boot sector and can be used to attack full disk encryption systems. Example bootkit is the "evil maid attack", which is a scenario where someone installs a bootkit on a device and mainly used to capture passwords for full disk encryption systems and other third-party security tools (Wilkins and Richardson 2013)
- **Hypervisor level:** The hypervisor Rootkits runs under the operating system kernel (Wang and Ghosh 2010). This category of Rootkits hosts the target operating system as a virtual machine, allowing it to intercept hardware calls made by the original operating system (Myers and Youndt 2007).

- **Firmware and hardware:** A firmware Rootkit uses a firmware to create a malware image in hardware and hides inside the firmware since firmware are not usually checked for code integrity. Firmware Rootkits are very difficult to remove and in most cases infected hardware requires replacement (Dufel et al. 2014). A hardware infected with a firmware Rootkit may allow the Rootkit to spread to other devices via peer to peer connections (Zhang et al. 2014).

In general, it is difficult to detect and find a Rootkit on your device because most of them are piggybacking with software you trust. No commercial software application exists for detecting Rootkits, however, there are different approach to follow for detecting Rootkits including behavioral-based methods, signature scanning, and memory dump analysis. In some cases, the only way for removing a Rootkit is to rebuild the compromised system. Other than that, protection mechanisms such as installing patches, updating applications and anti-virus software, discard emails with suspicious attachments or links, and be careful when installing new applications.

Ransomware

Ransomware is a type of malicious software that infects devices, blocks access to data – usually through encryption – and threatens to delete the data or publish it in case the victim did not pay the ransom – an amount of money – which is varies between $200 and $400. In most cases, Ransomwares are simple meaning that a knowledgeable person can easily reverse the effect of the Ransomware. On the other hand, some Ransomwares are very complex and causes a lot of losses to companies. Over the past 3 years, Ransomware has become one of the biggest cyber scams to hit businesses (Brewer 2016).

There are two main forms of Ransomware (Savage et al. 2015):

- Locker Ransomware: These are computer lockers that denies access to devices
- Crypto Ransomware: These are data lockers that prevent access to data and files by using encryption as one if the widely used techniques

Ransomware attacks are based on Trojans that appear to be a legitimate file that the users download and install, as phishing email that contains a malicious attachment, through drive-by download, infected storage device, or compromised Website. In case of complex Ransomware, such as "WannaCry", Ransomware travels between devices without users' interaction. Regardless the way a device got infected with a Ransomware, the victim may receive a pop-up window or email asking for paying the ransom, by a specific date in order to decrypt the date, if not paid, the amount might increase or data may be lost. Figure 2.4 shows an example Ransomware pop-up message asking for money to decrypt the victim's file.

Fig. 2.4 An example ransomware pop-up message

Five phases of Ransomware (Brewer 2016)
Ransomwares attacks go throw five phases and understanding each phase can be very helpful in preventing and mitigating the attack. These phases are exploitation and infection, delivery and execution, back-up spoliation, file encryption and user notification and clean-up.

1. **Exploitation and infection**: Ransomware file needs to be executed in order for the attacker to take over the victim's device. In most cases, this occurs through phishing or exploitation tool that looks for a vulnerability in the system to spread the Ransomware. These tools and phishing emails targets outdated software applications and operating systems.

2. **Delivery and execution**: Once the exploitation process succeeds in obtaining access to the victim's device, the Ransomware will execute and delivered to the victim's device. Once the Ransomware is executed it will move into persistence mode. The Ransomware is delivered over an encrypted channel, which makes things difficult to recover, and persistence will allow the Ransomware to continue the encryption process from where it is left in case the user tried to stop the Ransomware. The Ransomware can pick up where it left off and continue to encrypt the system until it is completed.

3. **Back-up spoliation**: Once the Ransomware infected the victim's device, it targets the backup file on the infected device and delete them to prevent the victim from

restoring files and data from that backup. Ransomware attacks are very powerful to the point it can unlock any protection mechanism on backup files, by killing the process, and delete the files to make recovery impossible.

4. **File encryption**: After back-up spoliation completed, the Ransomware will "perform a secure key exchange with the command and control server", establish encryption key, and tag the victim machine with unique identifier.
5. **User notification and clean-up**: With back-up spoliation and encryption completed, the victim is presented with a pop-up or email with instructions about extortion and payment. In most cases, as presented in Fig. 2.4, the victim is given few days for paying the money, if not the amount increases, and at some point the data is deleted.

Users can protect themselves from Ransomwares in different ways:

* Ransomware infections can be devastating and in order to protect against Ransomware attacks and other types of cyber extortion, experts urge users to backup computing devices on a regular basis and employ a data backup and recovery plan. Users also advised to update operating system and software – including anti-virus software – on a regular basis.
* End users should be aware of clicking on links in emails from strangers or opening email attachments and victims should do all they can to avoid paying ransoms. Restrict users' ability to download and install software, enable macros from email attachments, and use application white-listing – a list of software application that users are allowed to download and install on devices.
* While Ransomware attacks may be nearly impossible to stop, there are important data protection measures individuals and organizations can take to insure that damage is minimal and recovery is a quick as possible. Strategies include compartmentalizing authentication systems and domains, keeping up-to-date storage snapshots outside the main storage pool and enforcing hard limits on who can access data and when access is permitted.
* Report infection on time, treat it seriously, ask for expert assistance as soon as possible, and follow the security practice guidelines.
* Finally, experts discourage individuals and organization from paying money in order to unlock a device or decrypt the data as paying the money is not guaranteed to do so.

Worms

A computer Worm is a standalone malicious software application that spread to other devices on a network by replicating itself with the purpose of spreading malicious code (Weaver et al. 2003). Worms spread using network connections, emails, infected Web pages, and instant messages. In most cases, Worm causes some losses to the infected devices and network; at least worms can consume the network bandwidth and slow down the traffic. In some cases, Worms can install backdoors on infected devices. Unlike viruses, Worms spread automatically, and they do not require the attackers' guidance in order to spread over the network and other devices.

Worms spread by exploiting vulnerabilities; as a result, users need to keep software and operating systems updated, apply patches, and have countermeasures in place such as security software. Worms can be activated in different ways, including human activation, human activity-based activation, scheduled process activation, and self-activation (Weaver et al. 2003). The human activation works by convincing a user to execute a copy of the worm by deceiving the user into believing that a file, link, or attachment is a legitimate one. For example, users are deceived into activating the worms by including messages such as "important file attached", "you device is under security threat, download the following anti-virus software". On the other hand, with human activity-based activation, Worms are activated when the user perform a specific activity, such as resetting the computer, log into an account, etc. In scheduled process activation, Worms are activated using the scheduled system processes, such as application that are scheduled to run and install updates and patches, or may be processes for backing up the system. Finally, self-activation is considered the fastest method for spreading a Worm, where Worms initiate the execution themselves. The execution starts as soon as it can locate a vulnerability in the system and send the exploitation code.

Worms need to discover machines in order to infect them. A number of techniques by which a Worm discover those machines exists, including scanning, externally generated target lists, internal target lists, passive techniques. Scanning works by examining a set of addresses in order to discover vulnerabilities. Scanning can be sequential or random. This is the simplest method for discovering machines and widely used techniques. The speed of the scanning process depends on the density of vulnerable machines and the design of the scanner. For these worms, the Worm's spread rate is proportional to the size of the vulnerable population. Another technique is pre-generated target lists, which works by using a predefined list of targets, called a "hit-list" of probable victims. The scanning process can be accelerated using a small hit-list, where a complete list can infect all targets rapidly. The biggest challenge for this technique is the generation of the hit-list itself, especially for a comprehensive list. In externally generated target lists technique, an external list of victims is maintained by a separate server. This technique is used to speed a Worm attacking web servers. With internal target lists, applications provide information about other hosts providing vulnerable services. Such target lists can be used to create topological Worms, where the Worm searches for local information to find new victims by trying to discover the local communication topology. Finally, passive worms do not look for victim machines. Instead, they either wait for potential victims to contact the Worm or rely on user behavior to discover new targets. This is one of slowest techniques at the same time they are highly stealthy (Weaver et al. 2003).

Types of computer Worms (Goertzel 2009):

- **E-mail worm**: E-mail Worms use infected email attachments in order to spread. A special type of e-mail Worms is mass-mailing Worm which is embedded in an e-mail attachment and need to be opened by the victim in order for the Worm to install and copy itself.

- **Instant messaging (IM) worm**: Instant messaging worms spread via infected attachments to a message or even a link that has been sent to the victim, which links to a malicious Web site from which the worm is downloaded.
- **Internet Relay Chat (IRC) worm**: Comparable to IM worms, but exploit IRC rather than IM channels
- **Web or Internet worm**: Worms spread via a connection to a Web page, File Transfer Protocol (FTP) site, or other Internet resource.
- **File-sharing or peer-to-peer (P2P) worm**: In this category, Worms copy themselves via shared file folder, and then use peer-to-peer mechanisms to announce its existence in hopes that other peer-to-peer users will download and execute it.
- **Warhol worm**: This is a theoretical worm that can spread across the Internet to infect all vulnerable machines within 15 min of release or activation.
- **Flash worm**: This is a theoretical worm that is capable to spread and infect all target vulnerable machines rapidly and within few seconds. Flash Worms are possible by utilizing different scanning techniques that are able to generate a list of target machines.
- **Swarm worm**: This is a Worm that is theoretically based on emergent intelligence, which allows the Worm to cooperate with a large number of other Worms in order to exhibit emergent swarm behavior. Swarm Worms behave same as biological swarms and exhibit a high degree of learning, communication, and intelligence (Osorio and Klopman 2006).

Propagation techniques (Li et al. 2008)
- **Self-Carried**: Part of the infection process, a Worm is self-carried transmit itself to the target machine. This occurs part of self-activating scanning or topological worms, as the act of transmitting the worm is part of the infection process.
- **Second Channel**: Worms propagate using a secondary communication channel in order to infect the target machine. The infection occurs when the target machine connects back to the attacker's machine using to download the Worm body.
- **Embedded**: Using normal communication channels, an embedded Worm can send s itself to the target machine, either, either appending to or replacing normal messages. As a result, the propagation does not appear as anomalous when viewed as a pattern of communication.

The payload of a Worm is limited by the goals of the attacker. As a result, different payloads are designed in different ways depending on the attacker. The followings are different types of payloads used in Worms attack (Moya 2008):

- **None/nonfunctional**: A nonexistent or nonfunctional payload.
- **Internet Remote Control**: Worm opens a trivial to-use privileged backdoor on victim machines, giving anyone with a web browser the ability to execute arbitrary code.
- **Spam-Relays**: Creates an open-mail relay for use by spammers.

- **HTML-Proxies**: Redirects Web requests to randomly selected proxy machines make it difficult for responders to shut down compromised Websites. Those Websites are then used for illegal activities.
- **Internet DOS**: Denial of Service (DOS) attack is another type of payload that either targeted at specific sites or retarget-able under the attacker's control.
- **Data Collection**: Mainly used to compromise devices mainly used to store and manipulate sensitive data.
- **Access for Sale**: This payload is simply an extension to remote control and data-collection payloads, which allows for remote access to paying customers.
- **Data Damage**: This category of payload contains time-delayed data erasers used to manipulate data on the infected machine
- **Physical-world Remote Control**: This type of payloads is mainly used to control physical-world objects as well as influence human actions via computer devices.
- **Physical-world DOS**: This type of payload works by denying service in the physical world.
- **Physical-world Reconnaissance**: This type of payloads allows Worm to conduct further reconnaissance for later, non-Internet based attacks.
- **Physical Damage**: This type is mainly used to damage the target machine.
- **Worm Maintenance**: The type of payload is used to maintain the worm itself. A controllable and updateable worm could take advantage of new exploit modules to increase its spread.

Users can protect themselves from Worms in different ways. Users should download files from trusted sources, scan all files using anti-virus software applications, update anti-virus software application on a regular basis, avoid emails and attachments from unknown senders, make sure that the firewall in on and working, and finally, make sure that security policy and guidelines are followed, such as sing strong passwords and changing it frequently.

Trojan Horses

Trojan horses are malicious computer application, that deceives users by looking like a legitimate software application, but written with the purpose of harming devices or steal information such as bank accounts, passwords, and personal data. In general, Trojan Horses do not replicate themselves like other malicious applications; however, they can serve as a gate for other self-replicating malicious application to infect the victim's device by performing malicious actions such as opening a security hole into the victim's device.

Types of computer Trojan Horses (Goertzel 2009):

One type of Trojan Horses is called backdoor Trojan which mainly acts as a remote utility that allows the attacker to control the infected machine using a remote host. Denial of service (DoS) Trojan is category that falls under backdoor Trojans which allows the attacker to carry out distributed denial of service (DDoS) attack of the Trojan infection spreads widely enough. Another subtype that allows the attacker to control the victim machine is called FTP Trojan which opens port 21 and allows connection to the victim machine via FTP.

A second type of Trojan Horses is called data-collecting Trojan which works by collecting information from the victim's machine and send it back to the attacker. Under this category, several subtypes exist such as spyware Trojan, keylogger, screen logger, password spyware Trojan, and notifier. Spyware Trojan extends spyware functionality besides passive monitoring to active collection of personal information; interference with user control of the computer; modify computer settings, changing home pages, and/or loss of functionality. Remember that not all spyware programs are Trojans. A keylogger is used to monitor data input via keyboard and forward the captured data to the attacker. Screenlogger is used to take snapshots of the victim's machine and send it back to the attacker. Password spyware Trojan steals infected machines' passwords. Finally, notifier is used to confirm that the machine is infected and sends information back to the attacker. Such information includes Internet protocol address, open port numbers, email address, etc.

A third category that falls under Trojan Horses is called downloader or dropper, which are mainly used to download and install software that launches different kinds of malware n the victim's machine. Security software disabler, rogue security software, and ArcBomb are all considered Trojan Horses subtypes that falls under this category. Security software disabler works by terminating your anti-virus application or even your firewall in order to scan the victim's machine for possible vulnerabilities and potential attacks. Rogue security software deceives the victim into being a legitimate anti-malware software but, in fact, they are not, and leaves the victim's machine vulnerable to threats. Finally, an ArcBomb is an archived file that sabotages the archive de-compressor used to open archive files. Once the file is opened the malicious code is executed and floods the victim's machine useless data and causes the machine resources to be consumed leading to system crash.

A fourth category that falls under Trojan Horses is the proxy Trojan which works by turning a victim device into a proxy server that operates on behalf of the attacker. This is category is difficult to trace since if a forensics expert tries to trace it will lead to the victim's machine not the attacker machine.

A fifth category that falls under Trojan Horses are the Rootkits which represent a collection of tools used by the attacker to gain access to the victim's machine. Different malware uses Rootkits to hide them. Most Rootkits are installed when the attacker gain a user-level access to the victim's machine, once they gain access, they can change settings and anything they want.

A sixth category that falls under Trojan Horses is bots. A bot is simply any malware that allows the attacker to gain access and control of the victim's machine. Any machine infected with a bit is called a zombie or drone. A botnet is simply a collection of zombies controlled by the attacker who is capable of awakening those zombies and direct them to perform actions such as spam and phishing.

In general, Trojan Horses infect devices through social engineering where the users are deceived into opening an email attachment or clicking a link, or maybe by drive-by-download using a pop-up window. Signs such as low performance and changed settings are indicators that a Trojan horse resides on the victim's device. Counter measures such as keeping the anti-virus application up to date, avoiding clicking or downloading unknown files from unknown sources, and scanning files can help protecting against Trojan Horses.

Backdoors

A Backdoor is simply a method for bypassing authentication and encryption in a computer system. (Gandhi and Thanjavur 2012). Usually Backdoors are used to gain access to the victim device. In most cases, a Backdoor comes part of a software that appears to the victim to be a legitimate one, or as a separate program. Backdoors are mainly used part of a Trojan horse to gain access to the target device.

Rogue Security Software

Rogue security software is a malicious software the deceives users into believing it is a useful software from a security perspective and misleading them that a virus infected the device and asks for money to download and install a software that is capable of removing the infection. Rogue security software is considered a form of Scareware because it fears the victims and at the same time a form of Ransomware because it asks for money to download a removal tool (FOSsi et al. 2009).

In general, the victim is tricked into installing a Rogue security software using pop-up window with a message indicating that the user's device is infected, in other cases they are installed via drive-by-download. Once the Rogue security software is installed it can do many things such as disabling users' privilege from uninstalling the installed software, crashing and rebooting the device, and modify the systems' registries and settings.

To protect yourself from Rogue security software, make sure not to click on suspicious links, pop-ups or even email attachments, make sure to read user agreement and policy associated with each software you install, seek security personal advice when it comes to downloading and installing any software, and finally, users should try not to fall scared about your system being infected.

Internet Bots

An Internet Bot, Web Robot, or Bot us a malicious software application that runs scripts, simple and structural tasks, over the Web on a rate faster than human being (Dunham and Melnick 2008). Internet Bot are widely used in Web crawling or spiders that automatically fetch and analyze Web content from Web servers. Internet Bots can be good (Spiders), other are malicious. Internet Bots take advantages of Web servers, as a result servers outline rules that governs the behaviors of Internet Bots, where Bots that do not adhere the servers' rules should be denied from accessing the server and removed from the victim Website.

Modern Botnets can perform several tasks to the criminal controlling the Botnet, ranging from single purpose Botnets to Botnets that performs several tasks at the same time. Botnets falls into six main categories including spam, financial theft, DDoS, dropper, click fraud, intelligence gathering, cyberwarfare, and others (Zadig and Tejay 2011). Spam is a widely used Botnet technique, with the purpose of sending and delivering unsolicited emails for different purposes such as selling product. Botnets are mainly responsible for the majority of today's spam. From a legal perspective, sending spam messages is a criminal violation in the United States, but not other countries. Financial Theft Botnets are mainly used to target financial information. Such Botnets aims at cracking information related to bank accounts and steals password from banks' Websites, credit cards, and other information. Stolen information are sold by the attacker to gain some financial gain, and in most cases such information is sold for few dollars. Distributed Denial of Service (DDoS) Bots work through a large number of compromised devices that are used to attack a particular Website or online service. In most cases, the target Website or service will end offline or down from the attack. Other Botnets include Dropper Botnets that works by installing other types of malware, Click Fraud Botnets that works once the victim clicks an online advertisement, and Intelligence Gathering or Cyberwarfare Botnets where the purpose is usually state-sponsored surveillance or destruction for political means.

Keyloggers

A Keylogger, keystroke logging, or keyboard capturing is a surveillance software application that keeps track of the user behavior by logging and recording the keyboard strokes and steal sensitive information such as username and passwords. In some cases, Keyloggers are used to study human behaviors while they are interacting with computers. Keyloggers can be hardware based or software based.

A hardware-based keylogger is a simple device that resides between the keyboard and the device to which the keyboard is connected. The simple device act a repository for gathering information typed on the keyboard and store it on the simple device storage. The problem with such hardware-based keylogger is that the attacker must physically present in order to get the simple device back and access the recorded information. In case of wireless keyboards, a wireless sniffer can work similarly by intercepting data, decrypt it, and store it (Fig. 2.5)

On the other hand, a software-based keylogger do not require and physical access to the victim device. A software-based keylogger works by installing a piece of software on the victim machine, keep tracks of what is being typed in the keyboard,

Fig. 2.5 Example
hardware-based keylogger
(https://www.smartz.com/
blog/2011/02/01/hardware-
based-keyloggers-making-
identity-theft-easier/)

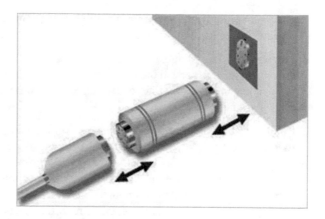

and then send the captured information to the attacker email or server. In general, software-based keylogger consists of two files, a dynamic link library (DLL) file that do the entire work, and another executable file that install the DLL file itself (Pathak et al. 2015)

Software-Based Keyloggers

From a technical perspective there are several categories of software-based keyloggers (Pathak et al. 2015):

- **Hypervisor-based**: theoretically, a keylogger can reside in a malware hypervisor underneath the operating system that becomes an effective virtual machine.
- **Kernel-based**: this category utilizes the root access to hide the keylogger in the operating system in order to intercepts keystrokes that passes the kernel itself. Since this is a kernel-based keylogger, it is hard to develop, write, and even to be detected by any protection mechanism. This category is considered a very powerful keyloggers, and in some cases can act as a keyboard device driver, which allows them to access any information typed on the keyboard as it goes to the operating system
- **API-based**: in this category, a keylogger hook a keyboard API in a running application and starts logging the keyboard strokes as if it a piece of software instead of malware. Windows APIs such as GetAsyncKeyState(), GetForegroundWindow(), etc. are used to poll the state of the keyboard.
- **Form grabbing based**: this category is based on Web form submission, where the keylogger records the information during the submission event. This is usually occurs when the victim completed the Web form and submit it over the Internet by clicking the button or hits the enter key.
- **Javascript-based**: this category is based on a malicious code that is injected into the target page, and works by recording all the information based on an event. Such code is injected using different methods such as cross-scripting, man-in-the-middle, or any other method
- **Memory injection based**: Memory Injection-based keyloggers works by altering the memory table associated with the Web browser. They works by injecting

a malicious code into the memory similar to malware techniques used to bypass Windows user account control.

Hardware-based keyloggers

From a hardware perspective there are several categories of hardware-based keyloggers. Such keylogger are not dependent on any software as they exist on the hardware level (Shi 2010).

- **Firmware-based**: This category is based on the computer BIOS that has been modified to handle keyboard events. In such category, the attacker requires a root level access to the victim machine in order to implement such keylogger.
- **Keyboard hardware**: in this category, a hardware circuit is attached between the computer keyboard and the computer. This circuit will be responsible for capturing the keyboard events. The events are recorded on an internal memory located on the circuit. The weakness of such keylogger category is that is that it can be easily detected. Also, it requires the attacker to be present in order to obtain access to the logged information on the circuit memory (Mali and Chapte 2014).
- **Wireless sniffers**: this category of keylogger works by sniffing keyboard events over a wireless connection. So, they are basically meant for logging wireless keyboard. For this method to work, the attacker needs to spend more effort cracking the encrypted connection between the wireless keyboard and the victim computer.
- **Keyboard overlays**: this category of keyloggers is widely used on ATM machine where fake keyboard is overlaid on the ATM in order to capture cards' PINs. The device is designed in a way that can easily deceive the victim into feeling that the overlaid keyboard is the real one.
- **Acoustic keyloggers**: this category of keyloggers is based on the sound created by each key on the keyboard. Each key on the keyboard creates a different acoustic signature when struck. Based on the acoustic sound, it is possible to relate the sounds to different keys (Kelly 2010).
- **Electromagnetic emissions**: this category works by capturing the electromagnetic emissions of a wired keyboard by tuning into a specific frequency of the emissions radiated from the keyboards and capture the information based on the emission (Vuagnoux and Pasini 2009).
- **Optical surveillance**: Optical surveillance is mainly used to capture passwords or PINs using a hidden camera, which allows the attacker to monitor the victims while they are entering passwords or PINs on the device.
- **Physical evidence**: This category utilizes physical evidence, such as fingerprints, in order to recover information. This category is widely used to crack a passcode on a smartphone or ATM machine, where the victim usually leaves fingerprints evidence that can be easily traced in most cases.
- **Smartphone sensors**: smartphones accelerometer can be used to capture keyboard strokes by placing the smartphone near the keyboard. The process works by detecting the keyboard vibration using the smartphone's accelerometer. Similar techniques can be utilized for capturing strokes on the touch-screen devices.

In some cases, detecting a Keylogger can be difficult. No single method is enough by itself. In general, anti-Keylogger application can help scan and lookup any installed Keylogger application. Application white-listing, a list of trusted application, can help prevent a Keylogger from being installed on the victim machine. Other methods include visual inspection to check if any suspicious device is connected to the machine other than regular peripherals. Post-prevention method, in case data has been compromised, includes using a 2-factor authentication in order to prevent the attacker for accessing any sensitive information. Finally, in some cases, on screen keyboard or voice to text technologies can reduce the chances data being captured by a Keylogger.

Crimeware

Crimeware is a category of Malware that is designed to facilitate illegal online activities and automate cybercrime such as identity theft or unauthorized access to the victim device. In some cases, a Crimeware steal sensitive information by installing a Keylogger on the victim device, which will capture and records sensitive information such as passwords and personal information. A complex Crimeware can be difficult to detect and prevent using Malware scanner.

Types of Crimeware (Goertzel 2009):
- **E-mail redirector**: E-mail redactors are used to intercept and relay outgoing emails to the attacker's system. Such redirector can be installed and run on a mail server, or even desktop computer.
- **IM redirector**: Similar to e-mail redirector, IM redirector works exactly the same by intercepting and relaying outgoing instant messages to the attacker's system.
- **Clicker**: A clicker works by sending commands to the victim's machine or replacing system files in which internet addresses are stored. These commands will cause redirecting the victim to a different Website or Internet resources.
- **Transaction generator**: This category targets a computer device that is located inside a transaction processing center, but not an end user device. They are mainly installed by hackers who targets transaction processing systems, such as credit card transaction data, to compromise its security by generating fraudulent transactions on behalf of the attacker
- **Session hijacker**: Session hijacking uses a malicious browser component to take over the victim's session after the victim logs in or starts a browser session. Session hijacking is meant to perform criminal activities such as money transfer from the victim's account.

Crimeware attacks go through the following cycle (Emigh 2006):

1. **Distribution**: Crimeware can be distributed in different ways depending on the Crimeware attack. Crimeware can be distributed via social engineering (for

example, piggyback attacks) or by exploiting a security vulnerability (for example, internet worms and hacking).

2. **Infection**: Crimeware infection takes many forms, however, in some cases, there is no infection stage in a Crimeware attack where the Crimeware itself is ephemeral. In cases where there is no infection stage, the Crimeware attack leaves behind no persistent executable code.

3. **Execution**: Crimeware execution occurs part of a one-time attack, as a background component of an attack, or by invocation of an infected component.

4. **Data Retrieval**: In this stage, the Crimeware retrieves the confidential data from storage, for example, in attacks such as data theft. In some cases data is retrieved and provided by the users themselves, for example, attacks such as keyloggers and web Trojans.

A Crimeware is distributed in many ways as follow (Emigh 2006):

- **Distribution via Attachment**: In this category, Crimeware is sent as an attachment (whether using email or instant messaging). Once received the victim is deceived into downloading and opening the attachment as it appears to be sent from a trusted source. Another approach for distributing Crimeware in this category us by embedding the Crimeware itself in a flash drive that usually left behind around the victim device. The victim who is finding the flash drive will attach the flash drive to their computer, which causes a malicious code to be executed automatically, or even install their own content.

- **Distribution via Piggybacking**: Crimeware can be distributed as part of an application that is downloaded for a different purpose. In most cases the downloaded application appears to be a legitimate applications and the Crimeware is part of it. This is a common mode of propagation for software that pops up advertising. In some cases, users' consent is obtained through end user license agreement in order o install the software as well as the Crimeware. This is common since users find end user agreement a lengthy and confusing process and difficult to understand, so the users simply agree the proceeds with the installation.

- **Distribution via Internet Worm**: Internet Worms that exploit security vulnerabilities can be used to spread Crimeware. This can be done by scanning an infected device and infect them with a Crimeware. Worms install a backdoor in most cases, which allows the attacker to use such backdoor to install a Crimeware on the target device. The infected device is used part of a botnet controlled by the attacker. In most cases, Crimeware installed vi backdoor includes keyloggers, phishing data collectors, and mail relays used for sending spam.

- **Distribution via Web Browser Exploit**: Crimeware can also be distributed via vulnerabilities in Web browsers. This usually occurs when the user visits a malicious Web site and a vulnerability (scripting, parsing, processing and displaying content, or any other component that can cause the browser to execute malicious code) is exploited by code on the site. In some cases, Crimeware payload is distributed via injection attack such as cross-site scripting.

- **Distribution via Hacking**: Crimeware can be distributed by hacking into the victim device by exploiting an existing vulnerability in the system. This is the most visible method of distribution in case the attack targets few devices.
- **Distribution via Affiliate Marketing**: offering financial rewards via affiliate marketing programs can be used to install Malware as well as propagating Crimeware through such affiliate networks. In most cases installed malicious software on a victim's devices are compensated with small payments and depending on the country where the victim is located.

Scareware

Scareware is a type of Malware that utilize social engineering in order to trick the victims by causing fear and anxiety through some signs that deceive the victims that the device has been infected and they need to buy a specific software to remove the infection. In that sense, a Scareware is considered a special case on Rogue security software (Stone-GrOSs et al. 2013). In most cases, Scareware presents the victim with messages in pop-up Window indicating that something wrong is happening such as "your device is infected with a virus", and trick the victim into buying a specific antivirus software application. Depending on the complexity of the Crimeware, the infection might be real or not, which mean that the minimum loss that can occurs to the victim is simply spending money on useless software. In some cases, the Crimeware pop-up message can lead to click-jacking and moving the victim to the attacker's Website or maybe initiates a Malware attack.

You can protect yourself from Scareware by following similar guidelines to other Malware including but not limited to using virus scanner and make sure that your anti-virus software is up to date, avoid clicking on pop-up Windows or malicious links in emails, close a pop-up Window using the "X" icon, and finally, if you think your device is infected do not panic and seek expert help immediately.

Viruses

A computer virus is a very sophisticated type of malicious software program that infect devices by modifying them in order to include an evolved version if itself. (Morales et al. 2006).

Viruses infect devices through deceiving users via social engineering or by exploiting security vulnerabilities in the target device. Viruses causes many different losses including but not limited too loss of sensitive information and waste computer resources.

Viruses are mainly classified into two types (Zuo and Zhou 2004):

- **Non-Resident Viruses**: Non-resident Viruses are executed with its host program, perform the intended function, infect the target files, and then transfers the control back to the program.
- **Resident Viruses**: Resident Viruses are activated whenever the user runs the infected program. Once activated, it loads replication module into the memory and then transfer control to the program. This category employs some system calls in order to reside in memory and modify some system calls by which they infect other programs.

Classes of computer viruses
- **File Infectors**: The majority of computer viruses target files that contain applications. When the user runs the infected application, the virus will be executed and installed on the victim machine, and starts copying itself. In most cases, the user is unaware of the existence of the virus (Abdelazim and Wahba 2002).
- **Script virus**: A script virus is a subset of file viruses written in a specific scripting language that mainly used to infect other scripts or can be part of multi-component virus. Script viruses are able to infect other file formats, such as HTML, if the format allows the execution of scripts (Goertzel 2009).
- **Boot-sector viruses**: Boot-sector viruses are designed to resign on the part of drive and run once the computer boots. Once the virus is installed and executed, it can infect any part of the hard drive. This way the virus will keep loading into memory every time the device is booted (Abdelazim and Wahba 2002).
- **Macro viruses**: Macro viruses are designed to be independent of the target operating system and infect files that contain data rather than programs. For example, office application cam run scripts – called macros – used to automate actions and activities. In some cases, Macros themselves can be programmed to be viruses and infect the target device by deceiving users into downloading and installing them as a legitimate macro (Abdelazim and Wahba 2002).
- **Electronic mail (email) virus**: An email virus is a kind of macro viruses. This type of viruses exploits weakness and vulnerabilities in the email application and spreads through the emails. E-mails are used to transmit any virus by copying and emailing itself to the contact addresses in the victim's email list. In some cases, they are sent as an attachment. The process proceeds into another cycle when the victim receive an infected email and open it (Goertzel 2009).
- **Multi-variant virus**: This type is the exact same core virus but with modification. In this case, the variant as well as other variants of the core virus will not be captured by the anti-virus software (Goertzel 2009).
- **Radio Frequency Identification (RFID) virus**: A theoretical virus that mainly target radio frequency identification devices (Goertzel 2009).

Viruses consist of three main parts (William 2008)
- **Infection mechanism**: Describe the spreading mechanism of the virus, the process of searching and locating target files on the victim machine.
- **Trigger**: Every virus has a trigger that activates the virus. Activation can be time based like a specific date, or activity based like clicking a link or file or opening a file.

- **Payload**: Payload refers to the malicious code that performs the virus routine. In some cases, the payload can be harmful by itself, or in other cases it can be distributive.

Virus life cycle (Subramanya and Lakshminarasimhan 2001) consists of four phases as follow
1. **Dormant phase**: The virus in the dormant phase is usually idle until it is activated, which means the virus cannot take any action in this phase
2. **Propagation phase**: In the propagation phase, the virus starts replicating itself into other systems or programs. In this phase, each system will contain an exact copy from the virus, which means it can go into another cycle of propagation. In some cases, the virus will change itself to evade detection.
3. **Triggering phase**. In the triggering phase, the virus is activated and starts the intended attack. The triggering of the virus can occur by different events, or based on time.
4. **Execution phase**: In the execution phase, the actual functionality of the virus is performed. In most cases, the functionality will harm the infected device by corrupting data or crashing the system.

Virus Detection Techniques Two main types of virus detectors exist. Signature based and behavior based detectors. Signature based detectors work by searching through objects for a specific sequence of bytes that uniquely identify a specific version of a virus. Behavior based detectors identify an object as being viral or not by scrutinizing the execution behavior of a program (Morales et al. 2006).

Countermeasures Many countermeasures exist to protect from viruses. The one widely known and used is installing an anti-virus software application. The anti-virus application will keep monitoring the system and block any attempt to infect the device. In other cases, the anti-virus software can scan the device in order to detect viruses and delete them from the infected device. It is highly recommended to keep your anti-virus database up to date in order to recognize latest threats because viruses are developed on a daily basis by attackers. Also, it is always a best practice to apply latest updates and patches to software applications and operating systems. Avoid clicking suspicious links, email attachments, or install software from unknown sources. Keep regular backups of your system data and information since viruses usually target those data.

Bugs

Software Bugs are simply errors, flow, or fault in a computer program that causes the software to behave incorrectly or in some cases stop working (Farchi et al. 2012). Since software Bugs are related to computer programs, they usually occur during the implementation phase or even in the design phase of application development. Bugs pose security threats to devices as it enables attackers to obtain unauthorized privileges.

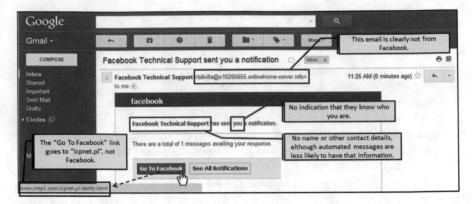

Fig. 2.6 Example phishing e-mail (https://blogs.otago.ac.nz/infosec/examples-of-phishing-emails/)

Phishing

Phishing, similar to using a bait to catch something, is considered a type of social engineering in which the attacker deceives the victim using spoofed emails to click a link or download and install a malicious application (Hong 2012) with the an attempt to gain unauthorized access to sensitive information. Figure 2.6 shows an example phishing email. You can easily detect a phishing email in case you are suspicious about it. First thing you notice about the example below is that it is from Facebook. However, if you examine different parts you can easily tell it a phishing email. First, the email address is not from the Facebook domain. Second, the technical support shows no identifiable information or contact details. Finally, before clicking any link inside the email, simply hover the mouse over it and you can easily notice the link details in the browser's status bar, which tells you the domain to which you are going to be redirected to.

Spamming

Spamming is the process of flooding the victim's device with unsolicited messages using email, social media, and instant messaging. The attackers usually send huge number of messages using a software that automatically creates accounts to be used to send spams. In general, spamming is a serious problem and no effective countermeasures exist to protect against spamming. Spam has variation including spim, spit, and spam SMS. Spim is specific to instant messaging (IM) applications. Spit or spam over IP uses the Internet to make telephone calls. Finally, spam SMS is a type if spam designed to be sent to mobile devices using SMS.

Browser Hijacking

Browser Hijacking is the process of changing the browser's default settings, such as the home page, without the user permission (Wang et al. 2015). In some cases, browser Hijacking can alter the browser aspects by adding extensions, toolbars, and bookmarks. Some attackers utilizing browser Hijacking can also include a Spyware part of the process.

Logic Bombs

A logic bomb is a piece of code that runs a malicious attack once it is triggered by a specific event (Rotich et al. 2014). One example of such may be triggered by an event can be a code that starts deleting files or encrypting files on the victim device. A time b is another variant of logic bomb that is triggered by time instead of an event.

Code Injection

Code injection is an attack that works by inserting a piece of code into an application from an outside source (Patel et al. 2011). Code injection can result in damages to valuable assets, leads to incorrect operations, or denial of service. In most cases, the process of code injection exploits weakness and vulnerabilities in the target system. Code injection takes many forms including but not limited to cross site scripting attack, SQL injection attack, and HTTP response splitting. SQL injection is widely used, when SQL injection based on 1 = 1 is always true (Fig. 2.7).

For example, suppose that you want to retrieve users' information based on UserID. In the application, nothing prevents the users from entering whatever information he/ she wants. I this case, the user enter some "smart" input like 223 OR 1 = 1. In this case, the SQL code looks like "SELECT * FROM Users WHERE UserID = 105 OR 1=1; ", which is a valid SQL statement and will retrieve all rows from the "Users" table since 1 = 1 is always true. Same statement can be used for different kind of purposes, especially retrieving sensitive information as follow "SELECT UserID,

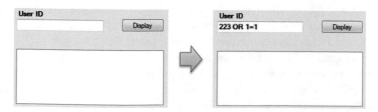

Fig. 2.7 A sample general form

Name, Password FROM Users WHERE UserID = 223 OR 1=1; ", which will allows the hacker to retrieve all users' accounts details from the Users table.

Types of injection attacks:

- **Code injection**: The process of injecting a malicious code or script into a program or Web application. In most cases this can execute an operating system command. In other cases, such code injection can allows the attacker to gain high privilege, which in turns might lead to full system compromise (Patel et al. 2011).
- **CRLF injection:** Carriage return line feeds (CRLF) Injection attacks works by injecting special characters into HTTP header values thus making possible to write arbitrary contents to the response body, including Cross-site Scripting (XSS) (Alenezi and Javed 2016).
- **Cross-site Scripting (XSS):** Cross-site Scripting works by executing a JavaScript code inside the victim's browser after being injected into a Website or Web application. The potential impact of such attack include account impersonation, defacement, and the ability to run arbitrary JavaScript in the victim's browser (Van Acker et al. 2012).
- **Email (Mail command/SMTP) injection**: This type of injection attempts to sends arbitrary emails by inserting data into the mail header using an input form. The attack occurs when injecting an IMAP/SMTP statement to a mail server that is not available using Web application. The potential impact of such attack include spam relay and information disclosure (Somani et al. 2012).
- **Host header injection**: Host header injection works by abusing the implicit trust of the HTTP Host Header in order to affect the password-reset functionality and web caches. Such injection attack can lead to password-reset poisoning and cache poisoning (Muscat 2017).
- **LDAP injection**: Lightweight Directory Access Protocol (LDAP) injection attacks use similar techniques to XPath Injection. They work by injecting an LDAP statement to execute an arbitrary LDAP command, which aims at extracting sensitive information, in order to gain access or modify the LDAP tree content. Such injection attack can lead to authentication bypass, privilege escalation, and information disclosure (Pérez et al. 2011).
- **OS Command injection**: This attack occurs when the attacker inserts an operating system command through incorrect user input that is passed to the system shell. This type of injection can lead to full system compromise (Liu 2015).
- **SQL injection (SQLi)**: Structured Query Language (SQL) injection attacks occur when the attacker injects an SQL statement that is parsed as part of the statements sent to the database. This type of injections can lead to authentication bypass, information disclosure, data loss/theft, loss of data integrity, Denial of service, and full system compromise (Buehrer et al. 2005).
- **XPath injection**: XPath injection attacks aims at exploiting a vulnerability in XPath expression used by an application and works by injecting data to execute a crafted XPath queries. This type of injection can lead to information disclosure and authentication bypass (Laranjeiro et al. 2009).
- **XML injection**: Extensible Markup Language (XML) injection is an emerging category of injection attacks that works by user's input s passed to an XML

document that is parsed by second-tier application or database, which in turns leads to injecting a malicious code into the database (Yee et al. 2007).

Pharming

Pharming is an attack in which the victim is redirected to a forged Website once he typed in the correct Website address (Nirmal et al. 2010). Pharming is done by exploiting a vulnerly in the domain name server, which are responsible resolving Internet addresses. Anti-virus software application are not capable of protecting against pharming, a special type of software called anti-pharming is required as a counter measure.

Figure 2.6 shows an example pharming attack. The attack is carried against www.vanguard.com. Initially, the pharmer arranges for the victim's DNS queries for www.vanguard.com to resolve to the pharmer's IP address, 6.6.6.6. Next, once the victim visits the Website, the pharmer returns a Trojan document containing malicious code referencing Website home page. The pharmer then updates the DS entry for the target Website to the address of Vanguard's legitimate server and denies subsequent connections from the victim. This causes the victim's browser to renew its DNS entry for vanguard, and load vanguard's legitimate home page. After the user authenticates him/herself, the malicious code in the Trojan document hijacks her session with the legitimate server (Karlof et al. 2007) (Fig. 2.8).

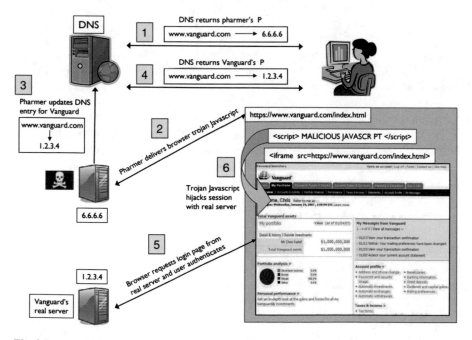

Fig. 2.8 Example pharming attack (Karlof et al. 2007)

Skills Section

General Malware Details

Visit the Avira virus lab https://www.avira.com/en/support-virus-lab and complete the following table:

Malware Type	Name	Impact	OS	Propagation Method	Brief Description
Adware					
Trojan Horse					
Malicious Application					
Worm					
Virus					
Backdoor					

Can you spot a fake email, message, or Website?
This exercise aims at determining whether you are skillful un telling the difference between a legitimate website/email/message and one that's a phishing attempt? Visit SocinWall phishing IQ test (https://www.sonicwall.com/en-us/phishing-iq-test-landing) or Opendns phishing quiz (https://www.opendns.com/phishing-quiz/) and complete the phishing quiz. Provide a screenshot of your quiz results.

Applications Section

Microsoft Safety Scanner

Visit **Microsoft** Website https://www.microsoft.com/en-us/wdsi/products/scanner# and download the Microsoft safety scanner. Microsoft safety scanner is a scan tool designed to find and remove malware from Windows computers. Simply download it and run a scan to find malware and try to reverse changes made by identified threats. Safety Scanner only scans when manually triggered and is available for use 10 days after being downloaded. We recommend that you always download the latest version of this tool before each scan.

How to run a scan

1. Download this tool and open it.
2. Select the type of scan you want run and start the scan.

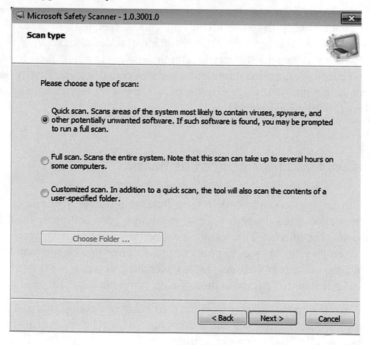

3. Review the scan results displayed on screen. The tool lists all identified malware.
4. Provide a Screenshot of your results

To remove this tool, delete the executable file (msert.exe by default).

White Hat: Experience With Keylogger

Visit CNET and download the free Wolfeye Keylogger. If you were not able to download it from CNET, simply search for it using Google search engine. The Keylogger runs secretly in the background and records every key that is pressed as well as the ability to make Screenshots on specified intervals. All the logged data can be sent to an email address which allows the attacker to monitor the victim's machine remotely.

1. Download the Wolfey Keylogger
2. Extract the installation file if necessary, otherwise simply run the application and install it on your device by following the instructions.
3. During the installation process, check the option under the Spy category
4. Adjust the screenshots and set the interval to 5 min. This should allow the application to send a screenshot every 5 min
5. Check stealth mode
6. Click Start.

Once you are done with setting up the application, spend few minutes performing different tasks such as entering text, send an email address, and open Web pages. Once one you are done with those tasks, run the keylogger application from background and check the captured information and associated folders. Briefly describe your experience with the keylogger application. Other than malicious attacks, how can we benefits from Keyloggers?

Protect Computers from Dangerous Website

Google Chrome is one of the widely used Internet Explorer. Chrome contains a feature, protection from dangerous sites, to protect your computer from threats while browsing the Internet. This feature is called Protection from Dangerous Sites. The feature employs techniques to protect from different threats and malware such as phishing and dangerous websites. In this exercise you will learn how to activate the protection feature in Chrome to protect your computer form different threats.

1. Start Chrome Internet browser from the start menu or desktop shortcut
2. Type the **chrome://settings/** in the address bar. You will be presented with a page that contains all Chrome browser settings

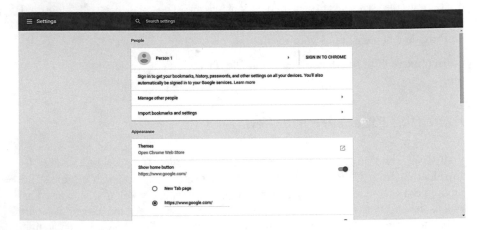

3. Proceed to **Privacy and Security**. Sometimes this area is hidden, just click on "Advanced" to expand
4. Make sure that "Protect you and your device from dangerous sites" is activated
5. Once you made the changes, restart the Chrome Browser

Limit Spam Messages in Your Outlook Account

1. Access your Outlook email account

2. From the ⚙ , click on Options. If you are not able to locate Options, simply use the search box and type Message Options.
3. From Message Options you can identify safe senders and receipts as well as blocked senders.
4. You can use the lists to control spam by allowing messages from specific senders or block messages from other senders and move them to Junk Email folder.
5. Under safe senders and recipients, enter an email address of a partner in the class and hit the enter key.

6. Under blocked senders, enter an email address of a different partner in the class and hit the enter key.

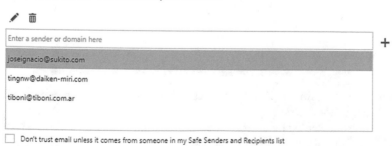

7. Ask both partners to send you an email message to the conFigured account. What happens when the messages arrive to you email? Where are these messages located in Outlook? Why?

Scan for Rootkits Using TDSSKiller

In this project, you download and install Kaspersky's TDSSKiller tool to help detect the presence of a Rootkit. TDSSKiller utility detects and removes malware such as malware family Rootkit.win32.tdss, bootkits, and Rootkits

1. Open your Web browser and enter the URL https://usa.kaspersky.com/downloads/tdsskiller
2. If you are no longer able to access the program through the above URL, then use a search engine like Google (www.google.com) and search for "Kaspersky TDSSKiller"
3. Click on Download Now and download the TDSSKiller compressed file
4. When the File Download dialog box appears, click Save and download the file to your machine
5. Extract the content of the compressed file to any location you want
6. Open the file and follow the instruction, after you accept the end user license agreement, to start TDSSKiller open the compressed (.ZIP) file.

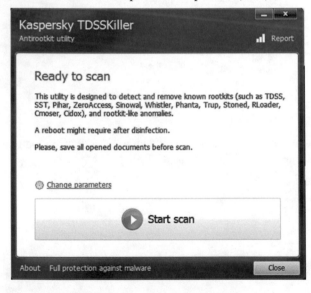

7. Before scanning for any Rootkit, you can change the scan parameter by clicking on Change Parameters
8. You will be presented by objects categories to scan as well as additional options.
9. Make sure that the first three objects categories are selected and click OK

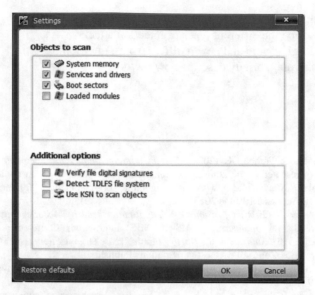

10. Start the scan process
11. Once the scan process completed, TDSSKiller will display a scan report.
12. Did you find any malware on your device? Provide a Screenshot of your scan result.

Questions
- Briefly explain what a Spyware is and how it affects devices.
- How can you protect against Spyware?
- What is the difference between Adware and Spyware?
- What is a Rootkit, briefly explain the different Rootkit categories?
- What is a Ransomware and how can you protect against it?
- Explain the five phases of a Ransomware.
- Briefly describe Worms, how they work, and how protect against them?
- Briefly discuss about Worms payloads.
- What is a Trojan horse?
- Briefly describe the types of Trojan Horses.
- What is the difference between Rogue Security Software and Internet Bots?
- What is a Keylogger?
- What is the difference between Crimeware, Scareware, and Ransomware?
- Explain the different types of Crimeware.
- Explain the life cycle of a Crimeware attack.
- Describe the ways through which Crimeware are distributed.
- What is a Virus, briefly describe the different Viruses categories?
- Describe the different parts of a Virus.
- Describe the Virus life cycle.
- What is the difference between Bugs and Backdoors?
- What is the difference between Phishing and Spamming?

- Explain what a Code Injection is and how it works.
- Briefly describe the different types of injection attacks.
- What is a Logic Bomb and how it differs from Time Bomb?
- What is Pharming, briefly describe how it works?

References

Abdelazim, H. Y., & Wahba, K. (2002). *System dynamic model for computer virus preva-lance*. Paper presented at the 20th international conference of the system dynamics society, Palermo, Italy, July, available at: www.systemdynamics.org/conferences/2002/proceed/papers/Abdelaz1.pdf. Accessed 19 June 2013.

Alenezi, M., & Javed, Y. (2016). *Open source web application security: A static analysis approach.* Paper presented at the engineering & MIS (ICEMIS), International Conference on.

Berberick, D. A. (2016). *Analysis of the North Atlantic Treaty Organization's (NATO) reaction to cyber threat.* Utica College\ProQuest Dissertations Publishing.

Brewer, R. (2016). Ransomware attacks: Detection, prevention and cure. *Network Security, 2016*(9), 5–9.

Buehrer, G., Weide, B. W., & Sivilotti, P. A. (2005). *Using parse tree validation to prevent SQL injection attacks.* Paper presented at the proceedings of the 5th international workshop on software engineering and middleware.

Chien, E. (2005). *Techniques of adware and spyware.* Paper presented at the the proceedings of the fifteenth virus bulletin conference, Dublin Ireland.

Dufel, M., Subramanium, V., & Chowdhury, M. (2014). Delivery of authentication information to a RESTful service using token validation scheme: Google Patents.

Dunham, K., & Melnick, J. (2008). *Malicious bots: An inside look into the cyber-criminal underground of the internet.* Boca Raton: CrC Press.

Emigh, A. (2006). The crimeware landscape: Malware, phishing, identity theft and beyond. *Journal of Digital Forensic Practice, 1*(3), 245–260.

Farchi, E., Raz-Pelleg, O., & Ronen, A. (2012). Software bug predicting: Google Patents.

FOSsi, M., Turner, D., Johnson, E., Mack, T., Adams, T., Blackbird, J., .Low, M., McKinney, D., Dacier, M., Keromytis, A., Leita, C. (2009). Symantec report on rogue security software. Whitepaper, Symantec, October.

Gandhi, V. K., & Thanjavur, T. N. S. I. (2012). An overview study on cyber crimes in internet. *Journal of Information Engineering and Applications, 2*(1), 1–5.

Goertzel, K. M. (2009). Tools Report on Anti-Malware. Retrieved from https://www.csiac.org/wp-content/uploads/2016/02/malware.pdf

Gordon, S. (2005). Fighting spyware and adware in the enterprise. *Information Systems Security, 14*(3), 14–17.

Gralla, P. (2005). *PC Pest Control: Protect your computers from malicious internet invaders.* Sebastopol, CA: " O'Reilly Media, Inc.".

Grégio, A. R. A., Jino, M., & de Geus, P. L. (2012). *Malware Behavior.* PhD thesis, University of Campinas (UNICAMP), Campinas

Hasan, M. I., & Prajapati, N. B. (2009). *An attack vector for deception through persuasion used by hackers and crakers.* Paper presented at the Networks and Communications, 2009. NETCOM'09. First International Conference on.

Hong, J. (2012). The state of phishing attacks. *Communications of the ACM, 55*(1), 74–81.

http://sarah-michelle-gellar.org/

https://www.pinterest.com/pin/194288171397349001/

https://www.smartz.com/blog/2011/02/01/hardware-based-keyloggers-making-identity-theft-easier/

https://blogs.otago.ac.nz/infosec/examples-of-phishing-emails/

Kapoor, A., & Sallam, A. (2007). Rootkits part 2: A technical primer. Retrieved from https://www. infopoint-security.de/open_downloads/alt/McAfee_wp_rootkits_part2_engl.pdf

Karlof, C., Shankar, U., Tygar, J. D., & Wagner, D. (2007). *Dynamic pharming attacks and locked same-origin policies for web browsers*. Paper presented at the proceedings of the 14th ACM conference on computer and communications security.

Kelly, A. (2010). *Cracking passwords using keyboard acoustics and language modeling*. Edinburgh: University of Edinburgh.

Laranjeiro, N., Vieira, M., & Madeira, H. (2009). *Protecting database centric web services against SQL/XPath injection attacks*. Paper presented at the database and expert systems applications.

Lemonnier, J. (2015). What Is Adware & How Do I Get Rid of It? Retrieved from http://www.avg.com/en/signal/what-is-adware

Levow, Z., & Drako, D. (2005). Divided encryption connections to provide network traffic security: Google Patents.

Li, P., Salour, M., & Su, X. (2008). A survey of internet worm detection and containment. *IEEE Communication Surveys and Tutorials, 10*(1). https://www.google.com/patents/US9158922

Liu, J. (2015). Method, system, and computer-readable medium for automatically mitigating vulnerabilities in source code: Google Patents.

Mali, Y., & Chapte, V. (2014). Grid based authentication system. *International Journal, 2*(10). http://www.ijarcsms.com/docs/paper/volume2/issue10/V2I10-0048.pdf

Medley, D. P. (2007). Virtualization technology applied to rootkit defense. Retrieved from http://dtic.mil/dtic/tr/fulltext/u2/a469494.pdf

Morales, J. A., Clarke, P. J., Deng, Y., & Golam Kibria, B. (2006). Testing and evaluating virus detectors for handheld devices. *Journal in Computer Virology, 2*(2), 135–147.

Moya, M. A. C. (2008). *Analysis and evaluation of the snort and bro network intrusion detection systems*. Intrusion Detection System\Universidad Pontificia Comillas. http://citeseerx.ist.psu.edu/viewdoc/download?doi=10.1.1.462.969&rep=rep1&type=pdf

Muscat, I. (2017). What are injection attacks? Retrieved from https://www.acunetix.com/blog/articles/injection-attacks/

Muttik, I. (2014). Preventing attacks on devices with multiple CPUs: Google patents.

Myers, M., & Youndt, S. (2007). An introduction to hardware-assisted virtual machine (hvm) rootkits. Mega Security.

Nirmal, K., Ewards, S. V., & Geetha, K. (2010). *Maximizing online security by providing a 3 factor authentication system to counter-attack'Phishing'*. Paper presented at the Emerging Trends in Robotics and Communication Technologies (INTERACT), 2010 International Conference on.

Osorio, F. C. C., & Klopman, Z. (2006). *And you though you were safe after SLAMMER, not so, swarms not zombies present the greatest risk to our national internet infrastructure*. Paper presented at the Performance, Computing, and Communications Conference, 2006. IPCCC 2006. 25th IEEE International.

Patel, N., Mohammed, F., & Soni, S. (2011). SQL injection attacks: Techniques and protection mechanisms. *International Journal on Computer Science and Engineering, 3*(1), 199–203.

Pathak, N., Pawar, A., & Patil, B. (2015). A survey on keylogger: A malicious attack. *International Joural of Advanced Research in Computer Engineering and Technology*. http://ijarcet.org/wp-content/uploads/IJARCET-VOL-4-ISSUE-4-1465-1469.pdf

Pérez, P. M., Filipiak, J., & Sierra, J. M. (2011). LAPSE+ static analysis security software: Vulnerabilities detection in java EE applications. *Future Information Technology, 184*, 148–156.

Rotich, E. K., Metto, S., Siele, L., & Muketha, G. M. (2014). A survey on cybercrime perpetration and prevention: A review and model for cybercrime prevention. *European Journal of Science and Engineering, 2*(1), 13–28.

Savage, K., Coogan, P., & Lau, H. (2015). *The evolution of Ransomware*. Mountain View: Symantec.

Schmidt, M. B., Johnston, A. C., Arnett, K. P., Chen, J. Q., & Li, S. (2008). A cross-cultural comparison of US and Chinese computer security awareness. *Journal of Global Information Management, 16*(2), 91.

Shi, P. P. (2010). *Methods and techniques to protect against shoulder surfing and phishing attacks*. Concordia University\Master thesis, Ottawa. http://dmas.lab.mcgill.ca/fung/supervision.htm

Somani, G., Agarwal, A., & Ladha, S. (2012). *Overhead analysis of security primitives in cloud.* Paper presented at the cloud and services computing (ISCOS), 2012 international symposium on.

Sood, A. K., & Enbody, R. (2011). Chain exploitation—Social networks malware. *ISACA Journal, 1*, 31.

Stone-GrOSs, B., Abman, R., Kemmerer, R. A., Kruegel, C., Steigerwald, D. G., & Vigna, G. (2013). The underground economy of fake antivirus software. In *Economics of information security and privacy III* (pp. 55–78). New York: Springer.

Subramanya, S. R., & Lakshminarasimhan, N. (2001). Computer viruses. *IEEE Potentials, 20*(4), 16–19.

Van Acker, S., Nikiforakis, N., Desmet, L., Joosen, W., & Piessens, F. (2012). *FlashOver: Automated discovery of cross-site scripting vulnerabilities in rich internet applications.* Paper presented at the proceedings of the 7th ACM symposium on information, computer and communications security.

Vuagnoux, M., & Pasini, S. (2009). *Compromising electromagnetic emanations of wired and wireless keyboards.* Paper presented at the USENIX security symposium.

Wang, S., & Ghosh, A. (2010). *Hypercheck: A hardware-assisted integrity monitor.* Paper presented at the Recent Advances in Intrusion Detection.

Wang, Y. M., Roussev, R., Verbowski, C., Johnson, A., Wu, M. W., Huang, Y., & Kuo, S. Y. (2004). *Gatekeeper: Monitoring Auto-Start Extensibility Points (ASEPs) for spyware management.* Paper presented at the LISA.

Wang, J., Xue, Y., Liu, Y., & Tan, T. H. (2015). *JSDC: A hybrid approach for JavaScript malware detection and classification.* Paper presented at the proceedings of the 10th ACM symposium on information, computer and communications security.

Weaver, N., Paxson, V., Staniford, S., & Cunningham, R. (2003). *A taxonomy of computer worms.* Paper presented at the proceedings of the 2003 ACM workshop on rapid malcode.

Wilkins, R., & Richardson, B. (2013). *UEFI secure boot in modern computer security solutions.* Paper presented at the UEFI Forum.

William, S. (2008). *Computer security: Principles and practice.* New Jersey: Pearson Education India.

Yee, C. G., Shin, W. H., & Rao, G. (2007). *An adaptive intrusion detection and prevention (ID/IP) framework for web services.* Paper presented at the convergence information technology, 2007. International conference on.

Zadig, S. M., & Tejay, G. (2011). Emerging cybercrime trends: Legal, ethical, and practical issues. In *Investigating Cyber Law and Cyber Ethics: Issues, Impacts and Practices* (p. 37). IGI global.

Zhang, F., Wang, H., Leach, K., & Stavrou, A. (2014). *A framework to secure peripherals at runtime.* Paper presented at the ESORICS (1).

Zuo, Z., & Zhou, M. (2004). Some further theoretical results about computer viruses. *The Computer Journal, 47*(6), 627–633.

Chapter 3
Security and Access Controls: Lesson Plans

Competency: Learn major aspects of security access controls in operating systems.

Activities/indicators
- Study reading material provided by instructor related to major aspects of security access controls in operating systems.
- Complete successfully an assessment provided by instructor related to competency content.
- For mastering levels, more than 80% of assessment grades should be earned in no more than three trials.
- Assessment questions can be pulled from the end of the chapter questions or any relevant material.

Competency: Learn major aspects of security access controls in database management systems.

Activities/indicators
- Study reading material provided by instructor related to major aspects of security access controls in database management systems.
- Complete successfully an assessment provided by instructor related to competency content.
- For mastering levels, more than 80% of assessment grades should be earned in no more than three trials.
- Assessment questions can be pulled from the end of the chapter questions or any relevant material.

Competency: Learn major aspects of security access controls in websites and web-applications.

Activities/indicators
- Study reading material provided by instructor related to major aspects of security access controls in websites and web-applications.
- Complete successfully an assessment provided by instructor related to competency content.
- For mastering levels, more than 80% of assessment grades should be earned in no more than three trials.
- Assessment questions can be pulled from the end of the chapter questions or any relevant material.

Competency: Learn information and tools used to build a RBAC (rule-based access control) system

---

(continued)

© Springer International Publishing AG 2018
I. Alsmadi et al., *Practical Information Security*,
https://doi.org/10.1007/978-3-319-72119-4_3

Activities/indicators

- Study reading material provided by instructor related to RBAC (rule-based access control) systems.
- Complete successfully an assessment provided by instructor related to competency content.
- For mastering levels, more than 80% of assessment grades should be earned in no more than three trials.
- Assessment questions can be pulled from the end of the chapter questions or any relevant material.

Competency: Learn information and tools used to build an OBAC (object-based access control) system

Activities/indicators

- Study reading material provided by instructor related to RBAC (rule-based access control) systems.
- Complete successfully an assessment provided by instructor related to competency content.
- For mastering levels, more than 80% of assessment grades should be earned in no more than three trials.
- Assessment questions can be pulled from the end of the chapter questions or any relevant material.

Competency: Learn how to create/modify access roles in different operating systems.

Activities/indicators

- Use one or more open source or free tools that allow users to create/modify access roles in different operating systems.
- For mastery, student is expected to try more than one operating system.
- For a mastering level, student should show what tool they selected, how they installed the tool and/or present a video to demonstrate the different commands they have used.

Competency: Learn how to create/modify access roles in different database management systems.

Activities/indicators

- Use one or more open source or free tools that allow users to create/modify access roles in different DBMSs.
- For mastery, student is expected to try more than one DBMS.
- For a mastering level, student should show what tool they selected, how they installed the tool and/or present a video to demonstrate the different commands they have used.

Competency: Learn how to create/modify access roles in different websites and web-applications.

Activities/indicators

- Use one or more open source or free tools that allow users to create/modify users and access roles in a website or a web server.
- For mastery, student is expected to try more than one website or web server.
- For a mastering level, student should show what tool they selected, how they installed the tool and/or present a video to demonstrate the different commands they have used.

Competency: Learn how to create/modify access roles in different tools used for RBAC (rule-based access control) system

Activities/indicators

- Use one or more open source or free tools that allow users to create/modify roles in an RBAC-based access control system (this can be related to one of the previous competencies, OSs, DBMS, websites, etc.). Instructor can make such judgment.
- For a mastering level, student is expected to show more than one RBAC example or application.
- For a mastering level, student should show what tool they selected, how they installed the tool and present a video to demonstrate the different commands they have used.

(Continued)

Competency: Learn how to create/modify access roles in different tools used for OBAC (object-based access control) system

Activities/indicators

- Use one or more open source or free tools that allow users to create/modify roles in an OBAC-based access control system (this can be related to one of the previous competencies, OSs, DBMS, websites, etc.). Instructor can make such judgment.
- For a mastering level, student is expected to show more than one OBAC example or application.
- For a mastering level, student should show what tool they selected, how they installed the tool and present a video to demonstrate the different commands they have used.

Competency: Be able to make an assessment (manually or through tools) for an existing access-control system for a particular information system or one of its components.

Activities/indicators

- Students should select a case study from their work environment or any simulated one to show current existing security and privacy policies.
- Students should present a critique and evaluation of those policies based on a base-line or a national standard.
- Students should also propose on what policies should be changed/updated or added for existing business policies to accommodate most recent standards.

Competency: Propose, design and implement a new security access control system for a particular information system or one of its components.

Activities/indicators

- Students should select a case study from their work environment or any simulated one to show current existing security and privacy policies.
- Students should present a critique and evaluation of those policies based on a base-line or a national standard.
- Students should also propose on what policies should be changed/updated or added for existing business policies to accommodate most recent standards.

Overview

Access controls are considered as important security mechanisms. They usually target (authenticated users: Those users who can legally access subject information system or resource). This indicates that they typically come after an initial stage called (authentication). In authentication, the main goal is to decide whether a subject user, traffic or request can be authenticated to access the information resource or not. As such authentication security control decision or output is a binary of either, yes (authenticated; pass-in), or no (unauthenticated; block). Access control or authorization is then considered the second stage in this layered security control mechanism. For example, it is important to decide whether subject user has a view/read, modify, execute, etc. type of permission or privilege on subject information resource. In this chapter, we will cover issues related to access controls in operating systems, databases, websites, etc.

Knowledge Sections

In the knowledge section, we will cover the role of access controls in operating systems, DBMSs and web applications. Those three categories represent the most mature information systems. Access controls exists in other types of information system with similar concepts and different types of implementations based on the maturity and the complexity of the information system. We will also study the two most popular access control architectures or models used to develop access control systems, RBAC and OBAC. In addition to OBAC and RBAC some information systems utilize other models. For example, Mandatory Access Controls: MAC (such as Biba and Bell-LaPadula) where control is centralized and managed by system owners. In contrary with the centrality nature of MAC, Discretionary Access Control (DAC) is un-centralized and users can manage access controls on resources they own. One example of DAC model is NTFS permissions on Windows OS. On NTFS each file and folder has an owner. The owner can use Access Control Lists (ACL) and decide which users or group of users have access to the file or folder. Most today's operating systems use DAC as their main access control model.

Permissions Permissions are the roles in an access control system to decide system constraints. The 3 main components in any permission include:

- **Entity**: The user/role or system that is granted/denied the permission.
- **Action**: This can typically include: view/read/write/create/insert/modify, etc.
- **Object**: This is the information system resource that will be the action object or where action is going to be implemented or executed.
    Those are the minimum three elements that access control permission should have. Based on the nature of the access control system, many other optional elements can be included.

Access Controls in Operating and File Systems

Operating systems evolve and continue to evolve with the continuous progress that we see in both the hardware and software industries. The operating system represents a complex software application that is used to manage all other software applications installed on that same personal computer, server, etc. It is also responsible to control and manage the communication between the three main entities in an information system: Users, software applications and hardware. The basic management modules that most mature operating systems include are: memory, processes, file, disk and network management. The tasks of access control in operating systems exist in different places and using different mechanisms. The basic structure depends on identifying users for the operating system and identifies their access levels on the different OS and system resources. In this sense, they combine access control with authentication. For example, users are prompted when they start an operating system to type a user name and password. Such user name and password should exist in the directory of authenticated users in the OS with the right password. User names and passwords fall in the category of authentication mechanism (something you know). Operating systems can also use other types of mechanisms such as: Something you have (e.g. an access card) or something you are (e.g. a finger print). System administrators can also decide different levels of constraints on the passwords that users are choosing for their accounts, how often they need to change it, etc. While user names and roles can be visible to other operating system users, passwords are encrypted. Rather than string the passwords themselves, hash values of those passwords are stored in the (shadow) passwords' folder.

You will learn in skills section how to check locations of accounts/passwords in Linux and Windows. We will also evaluate tools/methods to crack such passwords. Password crackers typically used either dictionary or brute force methods to crack those passwords. To counter dictionary-based password crackers, users should avoid using words from the dictionaries as passwords. To counter brute-force methods, systems should block users after a certain number of password attempts.

Users can be distinguished by unique names or they can use (roles) or groups and categories of people. Figure 3.1 shows a snapshot of Windows 10 computer management with built-in and generated user/role accounts.

Administrators (local host administrators) have the highest possible permissions in OS resources. A user can be added as an administrator when they are included in the (administrators group, Fig. 3.2).

In some operating systems, a special user (root) is defined as (a super user). Such uniquely identified user may have special privileges that cannot be granted to other created users or accounts.

There are basic system access control principles that designers/administrators should consider:

- **Principle of Least Privilege**: This means that the default access is nothing for a user. Users will then be granted access to the resources that they need. In this

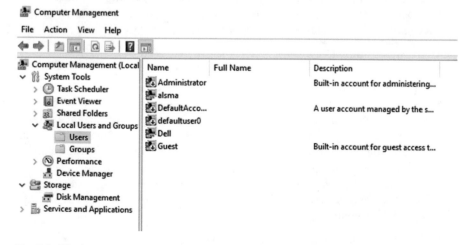

Fig. 3.1 Windows 10: Users and groups

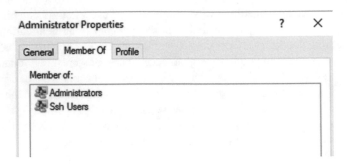

Fig. 3.2 Adding a user to the administrators group

regard, users are encouraged not to use administrator accounts all the time. If their accounts are exposed, with such high privileges, an attacker can significantly hurt the system. Alternatively, they should use normal or power user accounts and only elevate to administrators when they need to.

- **Separation of Duties**: Users should not accumulate responsibilities. They should not have open accounts that can play different roles in different occasions. Such roles should be divided.
- **Principle of Least Knowledge**: Similar to the principle of least privilege, users do not need to see resources that they have no associated tasks with. Intentionally or unintentionally users can abuse their privileges. In some phishing pr social engineering types of attacks, those users can be victims and their accounts can be used without their consent knowledge to commit attacks on systems or expose their resources.

Operating systems logically control access through Access Control Lists (ACLs). ACLs are mechanisms or concrete implementations of access control models.

ACLs represent permissions on system objects to decide who can have view/create/modify/execute a system resource or object. In operating system ACLs, an access control entry (ACE) is configured using four parameters:

- A security identifier (SID)
- An access mask
- A flag about operations that can be performed on the object.
- A flag to determine inherited permissions of the object.
- **Access controls in file systems:** File systems represent the management modules of files/folders in operating systems. As such, they inherent most of their access control roles from underlying operating system. File systems differentiate between user generated file and operating system file. Access control decisions on operating system files are usually decided by the operating system access control. They may need special/administrative level permissions before users can change their attributes.

The concept of file/folder ownership evolves when operating systems start to allow more than one user to access/use the same operating system environment and applications. With this evolution also, operating systems now have the ability to audit files history to see who did what and when. Those are usually critical information components to know when conducting a computer or digital investigation.

In Microsoft and Windows operating systems, one of the main goals of moving from FAT to NTFS file systems was to enable features related to file/folder access control auditing. FAT system was initially proposed before the era of operating system multi-users. In comparison with access control in database management systems access control in file systems focus on the file level access control. For a database, this will include for example roles on tables in comparison with roles on database schema in database access control.

UNIX file access permissions decide access on files based on three classes of users: Files' owners, members of the group which owns the file and all other user. Each of these three categories of users has permissions for reading, writing and/or executing.

Access Controls in Database Management Systems

Most of access control models and methods described in operating system section are applicable for Database management systems. Access controls restrict which entity can add/delete, modify or view an information resource. They can restrict access to specific attributes, specific tables or the whole database.

Unlike primitive flat-files based types of databases (e.g. MS Access) where users have exclusive access control on file-based databases, relational, object-based or object-oriented databases have more complex access controls that allow many users to have different levels of access controls on the same database. Shared types of databases may have access-control related problems from both authorized and unauthorized users. Security threats such as privilege escalation may cause some authorizations users to illegally have access resources privilege or permission they are not supposed to have.

- **User Views**: Each user in the database can have his/her own unique view of the database. In centralized databases, database administrators are expected to create and manage the different users' views and permissions. In the scope of "Big Data", a user view can be extracted from more than one database or data source and can include metadata customized based on the user preferences. User views can be also distinguished from authority levels. For example, in a University database, faculties, employees and students may all have "view" access to the students-grades table. However, in addition to seeing or viewing, faculties can create/modify/delete records in their own students' records. Some employees can have view/read/print access on those records. Students have also view/read/print access only on their own records from this, possibly large students-grades table.
- Database Authorization Table: Similar to Access Control Lists (ACL) in operating systems, database authorization tables contain users and their actions' limits. Table 3.1 below shows an example of a database authorization table for one employee or role of employees and a database system.

Table 3.1 A sample database authorization table

Authority level	Orders	Items	Payments
Read	Y	Y	Y
Insert/create	Y	N	Y
Modify	N	N	N
Delete	Y	N	Y

Table 3.1 indicates that in a large database management system, we may have a large number or instances of this authorization table

- **SQL GRANT/REVOKE commands**: Database administrators and privileged users can also change access control roles using user-defined procedures. For example, SQL GRANT command can be used to allow user(s) to have certain access levels on certain objects or resources. The general syntax for the command is:

  GRANT INSERT ON Students to Adam
  GRANT SELECT ON Students to students
  GRANT INSERT ON Students to Admin WITH GRANT OPTION

The addition (WITH GRANT OPTION) allows granted user/role to pass grants to other users/roles.

REVOKE statement is used to remove privileges from a specific user or role. The syntax is very similar to GRANT statement. Following are some examples:

  REVOKE DELETE ON students FROM Admin
  REVOKE ALL ON students FROM Adam

- **Data Encryption**: In addition to the different view/access level described earlier, encryption can be used to extend those levels. If an object or a resource is encrypted for a possible user, they may possible be able to view it, but due to encryption viewed data will not be human readable. In particular, encryption is used to protect data in-transit. This became more important recently with the cloud environment, where the whole database can be hosted remotely in the cloud and hence every single query will be transmitted over the Internet.
- **Inference Controls**: Research in statistics with information about humans exists in many disciplines. The significant growth of Online Social Networks (OSNs) in particular provides a wealth of information for statistical and social students. One problem associated with research or surveys in those areas is how to handle privacy issues on users' data. The goal of inference controls in data is to modify data to hide personal private information while not impacting data accuracy (Domingo-Ferrer and Josep 2009).

Access Controls in Websites and Web-Applications

Access control mechanisms in web systems are very critical to the security of those systems. When we survey the types and nature of web systems' attacks, we can see that the majority of those attacks target vulnerabilities/weaknesses in access control mechanisms (OWASP [3]).

While most access control models and methods n websites and web applications are similar to those on operating and database systems, however, they have their own unique access control aspects. For example, real time, current logged–in users, cache, cookies are all access-control web-related aspects that are unique in comparison with operating and database systems. On your own laptop, tablet or smart phone, you may not need to enter your credentials for your favorite websites (e.g. Facebook, Twitter, Gmail, Amazon, etc.), whenever you logged in. Those are cached for you from previous access times. It is important in the web environment to balance between security and user convenience. In real time environments, user-views are not only controlled by access-control roles, but also by user history, profile, favorites, etc. For example, when a user logs-in to Amazon or e-bay, many of the items they will see (e.g. future possible items to buy), are customized based on the user profile and their buying history.

Most web systems include two main classes of users: Administrators to create and maintain web pages and users or customers to use those pages. Users in most cases are not allowed to create/delete or modify pages. As users, most of the data they communicate with exist in the backend databases.

Identity Management

In the Internet or web environment, many of the personal/physical types of identification methods are un-applicable or impractical. On the other hand, as e-commerce and the use of the Internet as a business media is continuously growing, identity theft is a very serious issue. Yearly reports on monetary loses due to identity theft are showing that such problems will continue to be serious security problems in the near future.

Identity management systems are information systems in charge of the source of authenticating e-commerce parties to each other. Identity management systems work as authentication, rather than authorization systems. Access control systems represent a second stage security layer after identity management in a layered security system. An access control system that cannot first properly distinguish authenticated from unauthenticated entities will fail in all access control tasks.

Single-Sign-On (SSO)

In addition, to caching users' access profiles on their laptops, smart phones, tablets, etc., websites offer SSO methods to allow users to access all website resources with only one time access or login request. Even if the website includes different backend databases and servers, user credentials will be passed on from the first page that requested the credentials to any other system that require those credentials.

Session Time-out

Users create or start sessions when they login to a website. If the website detects that user has been inactive for some time, the session will be closed and the user has to login again when they come back. There is no specific time that all websites use where after the session will expire. Rather, the session time-out depends on the time the user has been "inactive" since they started the session. Such "inactivity" can be observed from the website when the user is not triggering any object in the websites. If the user is reading some documents in the website without any mouse, keyboard, touch-screen interactions, he/she can be seen as "inactive" by the system.

There are four types of session-based attacks: interception/hijacking, prediction, brute-force, and fixation. Session time-out can be one of the effective methods to counter those types of attacks. Other methods include session encryption, long and randomly generating session IDs.

Sessions can be location, rather than time sensitive. For example, your account in anyone of those websites may stay open in your smart phone for a very long time. However, if you tried to access your account from a new machine or phone that you never used before, this will trigger a new session.

Most current websites try to combine more than one authentication method to counter username/password hackings. For example, websites such as: Gmail, Yahoo mail, Facebook, etc. encourage users to include their phone numbers or secondary email accounts. If a user tries to login to an account from a new machine (a machine that was not associated with this user before), the website may send to the phone or the secondary email a verification code. The user is then asked to provide such code before accessing the account.

Kerberos

Kerberos is a ticket-based network authentication protocol that allows nodes communicating over a non-secure network in order to prove their identity to one another. Kerberos is proposed as an alternative to logical or password-based authentications that can be defeated by eavesdropping. In addition, this can be a convenient option where users do not need to remember their user names and passwords. In Kerberos, Information for authentication is sent encrypted between communication parties. Current Kerberos uses the data encryption standard (DES) for encryption.

Digital Certificates

Software companies lost a significant portion of their forecasted profit due to illegal copy, download and transfer of software between users through the Internet. Digital certificates are proposed to authorize the users' downloaded copies of software applications. When users buy/download new commercial software, they are asked to activate their copies as soon as they are online. Typically, a digital certificate (in a file format) is created for the user machine and is transferred to software company license server. The certificate is embedded with information related to the hardware components of the user host (e.g. disk drivers, CPU, physical memory, network card, etc.). Users may need to reactivate their license if they change any of that equipment.

Certificates are also used in identity managements. They are typically used as alternatives to user credentials (i.e. login user names and passwords). Files with information about communication partners are saved locally. Information in those files is exchanged once the user logs in to the target server.

Access Control in Distributed and Operating Systems

Access Control Lists (ACL) in the most popular Network Operating Systems (NOS), NFS are inherited from UNIX-based operating systems. In some NOSs (e.g. Andrew file system, AFS), ACLs are based on directories rather than files. As an alternative to ACLs, some NOSs (e.g. WebFS) use authorization certificates or a hybrid scheme of both access control implementations. Certificates are created and authorized by third parties; Certificate Authorities (CA).

Network switches and routers include their own operating systems. Part of their operating systems, they also include access control systems or modules. In addition to access controls related to users, NOS may include access controls for other systems, applications and even traffic. Host-based access control uses host IP, DNS and possibly MAC addresses. Several network-based attacks related to those attributes are possible. Examples of those attacks include: IP/MAC/ARP spoofing, DNS pharming, etc. Figure 3.3 shows a sample ACL from a Cisco router.

```
access-list 111 deny ip host 0.0.0.0 any log
access-list 111 deny ip 127.0.0.0 0.255.255.255 any log
access-list 111 deny ip 10.0.0.0 0.255.255.255 any log
access-list 111 deny ip 172.16.0.0 0.15.255.255 any log
access-list 111 deny ip 192.168.0.0 0.0.255.255 any log
access-list 111 deny ip my.net.15.0 0.0.0.255 any log
access-list 111 permit tcp any host my.net.15.3 eq 22
access-list 111 permit tcp any host my.net.15.66 eq smtp
access-list 111 permit tcp any host my.net.15.66 eq 22
access-list 111 permit tcp any host my.net.15.66 eq www
access-list 111 permit tcp host 131.154.1.3 host my.net.15.3 eq domain
access-list 111 deny tcp any any eq domain log
access-list 111 permit udp any host my.net.15.3 eq domain
access-list 111 permit udp any eq domain any
```

Fig. 3.3 A sample router ACL

More recent software-controlled NOSs such as Software Defined Networking (SDN) Controllers aim at giving users and their applications a fine-grained level of access control on traffic from and to the network (For more details, see OpenDayLight project or platform at: https://www.opendaylight.org .

One use case which is related to access control and mentioned in SDN projects is related to wireless AAA (Authentication, Access control and Accountability). In our home wireless internet, those three are typically combined in one. This is since current methods assume one user account and management systems for the three functions (Alsmadi and Dianxiang 2015).

RBAC (Role-Based Access Control)

Role-based access control (RBAC) is a popular access control model for kernel security control enforcements. Many of the policy-based security systems adopt RBAC when writing, and enforcing security roles (NIST 2010 report). The report showed that RBAC adaptation continuously increases between the years 1992 to 2010. The NIST RBAC model is defined in terms of four model components: Core, Hierarchical, static separation of duty relations, and dynamic separation of duty relations.

RBAC is a policy-based centralized access control system. In terms of centralization, it is similar to Mandatory Access Control (MAC) where access to system resources is controlled by the operating system and system administrators. Unlike centralized control in MAC, Discretionary Access Control (DAC) allows users to control access to their own resources (only). In DAC, resource objects have ACLs associated with them.

Figure 3.4 shows the basic RBAC model. Users should not be given access control permissions based on their own identities. They should be assigned to "logical" roles. Eventually they will request to access resources based on their assigned roles. Permissions are assigned to roles. Permissions decide actions on

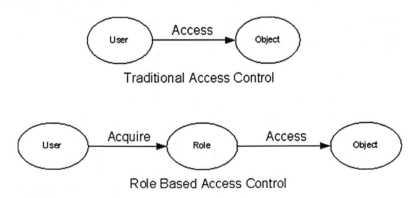

Fig. 3.4 Basic RBAC Model

objects or resources. This can facilitate security auditing for a large system. Roles in policies should not make references to users, their names, login names, etc. They should make references to the different types of roles employees play in the different information systems. Relations between users, roles, permissions and objects or resources are not exclusive. Different users can be assigned to the same role. Different roles can be given same permissions (some similar permissions, as if all permissions are identical then maybe those different roles should be combined).

Operations or sessions include activities or requests performed by users. Policies or roles in the access control system can decide whether to authorize those operations or not. Objects refer to system resources. The granularity level of such objects can vary significantly. For example, the whole DMBS can be seen as an object in one role. In another role, a table, a field, a record or even a cell of a database in that DBMS can be the object in the role request.

As the term implies (roles), RBAC consider the roles of users or entities as the basis to create policies. RBAC systems include a set of (Policies). A policy can be a high level (human nature) statement. For example, the following can represent a security policy (Users of the company network should not be allowed to access websites with improper contents). The followings can be stated about this policy:

- This policy can be easily understood by humans.
- This policy is very abstract in terms of network or security tools. In other words, no existing security tool (e.g. a firewall, a router, a Virtual Private Network (VPN), an Intrusion Detection or Protection System: IDS/IPS) can take this policy and enforce it (without extra tools/programs/humans to do the translation between this high level policy and low level security roles).
- Typically, such high level policy can be enforced using a set of low level roles that can be understood and enforced by security control systems.

In RBAC users/roles can't make changes on policies/roles that impact them. RBAC policies are usually created and managed by system administrators.

Users are different from roles in RBAC. Typically, roles are created and then each user can be assigned a role. For example, in a University system, we can define roles such as: Employee, Faculty, Student, System administrator, Manager, etc. Users can also play more than one role simultaneously or in different times, environments, functions, etc. For example, in a University system some employees can be students as well. This means that the RBAC system should be able to differentiate the current role they are playing while using or requesting a resource and hence assign them to the right role.

RBAC is popular in network-based information or distributed systems. This is largely as policies in network or security components are written as roles. For example, policies in firewalls (software and hardware), routers, switches, IDS/IPS, etc. are all written as roles.

Roles in RBAC can also have hierarchies where children roles can inherit permissions from parent roles. For example, in a network domain controller, domain controller administrators have the highest, or root permission level. This means that they can access resources in lower levels (e.g. host computers) without the need to be host administrators, power users or even users in those local hosts. Their root level role implies having an administrator privilege in all network or system resources.

OBAC (Object-Based Access Control)

With the evolution of object oriented paradigms in software design, programming and database management systems, OBAC evolves recently. In OBAC, objects rather than roles are the central entity in the access control systems. In OBAC, not only users, roles, or entities are objects, but also resources and decisions themselves can be objects (to have their own attributes and actions).

Attribute-based access control (ABAC) falls in the same category of OBAC in trying to design roles with finer details. In ABAC, roles include who (user-role) is going to do what (permission), on what (object or resources). However, each one of those entities (i.e. users, roles, permissions and objects) can have attributes and values for those attributes. Hence the final operation can be denied or permitted based on values of some attributes. For example, in the policy role example (Employees can use company Internet services within work-time or weekdays), has the same users, roles, permissions and objects. However, while all those are identical, in one instance or operation, such employee will be permitted (if it's Monday) or will be denied (if its Sunday). Similarly, one of the main drivers to use OBAC as a replacement to ABAC is that in OBAC we can make policy roles with much finder details in comparison with RBAC. Those roles can be also dynamic, rather than static. Attribute values for different entities can change with time, or some other environmental factors. Policy enforcement system is expected to be more complex to be able to evaluate the attribute values of all entities related to the operation or request.

Figure 3.5 shows OBAC/ABAC reference architecture. Major components are:

- Policy Enforcement Point (PEP): This module is in charge of enforcing the final (action decision) for an access control request. It will receive the request and context from the application. It will also receive final decision from Policy Decision Point (PDP).
- Policy Decision Point (PDP): This module represents the decision engine in the access control system. This is why it communicates with most components.
- Actual policies or instances of policies exist in policy store. Network administrators can have access to this store in order to create/delete/modify policies.
- Policy Information Points: In addition to policies, system should also know the different types of entities (i.e. users, roles, entities, objects, permissions). All those should exist in PIP. Each one of those can have its attributes. Attribute values can be extracted in real time from the application, environment, etc. Each operation or request can make queries to read a different set of entities or entity attributes.

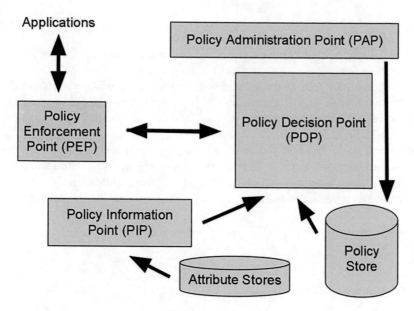

Fig. 3.5 OBAC/ABAC Reference Architecture

- Policy Administration Point (PAP): For general access control system control and management issues.

  As the most recent models of access control, in skills' section, you will be asked to evaluate a concrete OBAC/ABAC access control implementation.

There are some challenges that exist in all access control systems such as the nature of relations between different roles. For example, in roles' conflicts' cases, different roles in an access control system may contradict and negate each other actions or decisions. Final decision made on a request may not what system administrators are expecting. Roles can be also contained within other roles where inner roles are hidden and can never be enforced in any permission request. In such case, those roles, while they exist, they are dead or useless.

Skills Section

Tools for Windows SAM Database

- Windows SAM database typically includes important information.
- In this task, you are expected to run your own experience to show how can you extract (SAM Database information) from a subject or suspect machine.
- You can use your own machine for that case. Alternatively, if you don't want to use your own machine or don't have Windows, you may use any open source image (from our links or the Internet).

- Typically, SAM database exists in (Windows/system32/config/SAM). Of course that file (SAM) will not allow you to open it or even copy it (when OS is live or online), using normal Windows tasks, and hence you have to look for alternative tools to do that.
- Document which tool, you used and how you made a copy of your SAM database (don't work on the original one, if it's your live machine). Then also you should document, using the same tool, or another one, what information you can find in the SAM database.
- Your final goal is to see if you can crack the passwords for the users in the SAM database. Here are some tools that you may use: (samdump2 (https://source-forge.net/projects/ophcrack/files/samdump2/), fgdump (http://foofus.net/goons/fizzgig/fgdump/downloads.htm), regripper (https://code.google.com/archive/p/regripper/downloads).
- ---

Users Access Control in Linux

(Use your name/variables instead):

- (to add a user): sudo adduser izzat
- (to delete user): sudo userdel izzat
- (to delete user and the user home directory): sudo userdel -r izzat
- (to manually delete the user home directory if it is left behind): sudo rm. -r / home/izzat
- (all user information is stored in a text file /etc./passwd)(users passwords information file)
  (usernames and home directories), (to see all the information for users)
- (you can rename the username from this file): sudo vi /etc./passwd (or sudo nano /etc./passwd)
- (to change the user password): sudo passwd izzat
- (to clear the screen): clear
- (put users into groups and give permissions to groups)
  (add a group): sudo groupadd tamusa
  (delete a group): sudo groupdel tamusa
  (add a user in a group): sudo adduser izzat tamusa
  (to remove a user from a group): sudo deluser izzat tamusa
  sudo gpasswd -a izzat managers
  sudo gpasswd -d izzat managers
- (to see all the information about the groups)(groups configuration file)
- sudo vi /etc./group (sudo nano /etc./group)
- (to see all the groups you are a member of) groups (to see all the groups):
- sudo nano /etc./group
- (to change the permissions): sudo chmod 755 folder1 -R

- sudo chmod 775 file1
  (chmod stands for change mode), (−R switch is for a folder)
- to see the permissions for files and folders in the current directory- > ls -l
  drwxrwxrwx 2 root root test1, (rwxrwxrwx means 777 permission)
- sudo chmod 755 folder1 −R, 7 = rwx, 5 = rx-, ls -l
  drwxr-xr-x 2 izzat group1 test1
- sudo chmod u + w file1, sudo chmod g + r file1, sudo chmod o + x file1
- sudo chmod -R u + w folder1, sudo chmod u-w file1, sudo chmod g-r file1, sudo chmod o-x file1
- sudo chmod u + rwx,g + rw,o + x file1, sudo chmod u-rwx,g-rw,o-x file1
- sudo chown -R user1 folder1 (to change the user ownership of a folder and sub-folders)
- sudo chown user1 file1 (to change the user ownership of a file)
- sudo chgrp -R group1 folder1 (to change the group ownership of the folder and sub-folders)
  sudo chgrp group1 file1 (to change the group ownership of the file)
- sudo chgrp -R izzat folder1, drwxrwxrwx 2 izzat root 4096 test1
- sudo chgrp -R group1 folder1, drwxrwxrwx 2 izzat group 4096 test1
- (changing ownership in one line)(−R is for recursive in case of a folder)
  sudo chown izzat:tamusa file1, sudo chown -R izzat:tamusa folder1

Applications Section

You are expected to select an instance from a DBMS and conduct testing/auditing the following security-related components. Summarize your findings:

- Users/roles/permissions.
- Responsibility for authority tables and sub-schemas.
- Biometric controls (if any exists)
- Encryption controls (if any exists).

Access Controls in Different Websites and Web-Applications

You are expected to select an instance from a website or a web application and conduct testing/auditing the following security-related components. Summarize your findings:

- Users/roles/permissions.
- Responsibility for authority tables and sub-schemas.
- Biometric controls (if any exists)
- Encryption controls (if any exists).

Applications for RBAC (Role-Based Access Control) Systems

The majority of current access control systems are based on RBAC. Pick an information system and describe 5 instances of roles, specifying in each role: Users, permissions, and system resources.

If you are not sure if the system you are selecting is adopting RBAC, assume so.

Applications for OBAC (Object-Based Access Control) Systems

- XACML "eXtensible Access Control Markup Language" is the current de-factor standard implementation for ABAC or OBAC.
- Go to github.com (or any open source website). Search for one implementation for XACML in your preferred language. Then based on your programming skills, demonstrate and present the implementation in class and how it can be used.

Questions
- In the scope of access controls, describe the difference between permissions and privileges.
- In the scope of access controls, describe the difference between system objects and resources.
- In the core of access control, the concept of AAA (Authentication, Access control and Accountability). Describe the three terms and elaborate in their roles in an access control system.
- Make a comparison between the different access control systems: DAC, MAC, RBAC, ABAC and OBAC.
- Describe examples of threats that can target access control systems.
- What is the different between a centralized and non-centralized access control models?
- Users or software applications can both make access control requests to access system resources. Do access control systems differentiate between requests from users or software applications? Explain your answer.
- Why ABAC or OBAC are recently proposed as alternatives to RBAC? What are new advantages they can bring?

References

Alsmadi, I., & Dianxiang, X. (2015). Security of software defined networks: A survey. *Computers & Security, 53*, 79–108.

Domingo-Ferrer, Domingo-Ferrer J. (2009). Inference Control in Statistical Databases. In: LIU L., ÖZSU M.T. (eds) Encyclopedia of Database Systems. Springer, Boston, MA.

Li, N., Mao, Z., & Chen, H. (2007). Usable mandatory integrity protection for operating systems. In *Proceedings of IEEE symposium on security and privacy* (pp. 164–178). Berkeley, California: IEEE Computer Society Press.

Miltchev, S., et al. (2008). Decentralized access control in distributed file systems. *ACM Computing Surveys (CSUR), 40*(3), 10.

NIST. (2010). A report on: 2010 economic analysis of role-based access control. http://csrc.nist.gov/groups/SNS/rbac/documents/20101219_RBAC2_Final_Report.pdf.

OWASP. https://www.owasp.org/index.php/Access_Control_Cheat_Sheet#tab=Other_Cheatsheets.

Shaffer, M. (2000). Filesystem security – ext2 extended attributes [online]. Available from: http://www.securityfocus.com/infocus/1407.

Chapter 4
Security and Risk Management and Planning: Lesson Plans

Knowledge Activities

Competency: Define Risk

Activities/Indicators
- Study reading material provided by instructor related to major aspects of identifying risks.
- Complete successfully an assessment provided by instructor related to competency content.
- For mastering levels, more than 80% of assessment grades should be earned in no more than three trials.
- Assessment questions can be pulled from the end of the chapter questions or any relevant material.

Competency: Determine an organization's risk tolerance

Activities/Indicators
- Study reading material provided by instructor related to the types of risks tolerance used by organizations.
- Complete successfully an assignment provided by instructor related to the competency content.

Competency: Understand the process for managing security risk

Activities/Indicators
- Study reading material provided by instructor related to the process of managing risks.
- Complete successfully an assignment provided by instructor related to competency content.

Competency: Understand the impact of risks

Activities/Indicators
- Study reading material provided by instructor related to the impacts of risks.
- Complete successfully an assessment provided by instructor related to competency content.
- For mastering levels, more than 80% of assessment grades should be earned in no more than three trials.
- Assessment questions can be pulled from the end of the chapter questions or any relevant material.

(continued)

© Springer International Publishing AG 2018
I. Alsmadi et al., *Practical Information Security*,
https://doi.org/10.1007/978-3-319-72119-4_4

Competency: Discuss why disaster recovery planning is necessary

Activities/Indicators
- Study reading material provided by instructor related to the importance of disaster recovery planning.
- Complete successfully an assessment provided by instructor related to competency content.
- For mastering levels, more than 80% of assessment grades should be earned in no more than three trials.
- Assessment questions can be pulled from the end of the chapter questions or any relevant material.

Competency: Identify vulnerability threats

Activities/Indicators
- Study reading material provided by instructor related to the tools and techniques used for threat and risk identification.
- Complete successfully an assignment provided by instructor related to competency content.

Competency: Understand risk response

Activities/Indicators
- Use one or more templates to outline a risk response strategy.
- Complete successfully an assignment provided by instructor related to competency content.

Competency: Discuss tools for tracking risks.

Activities/Indicators
- Use one or more risk tracking tools to manage and track risks.
- For a mastering level, student should show what tool they selected, how they installed the tool and/or present a video to demonstrate the different features they have used.

Competency: Understand risk monitoring

Activities/Indicators
- Study reading material provided by instructor related the process of monitoring risks.
- Complete successfully an assignment provided by instructor related to competency content.

Competency: Identify risk metrics

Activities/Indicators
- Study reading material provided by instructor related to the techniques used for identifying risk metrics.
- Complete successfully an assignment provided by instructor related to competency content.

Overview

This chapter covers risk management and planning associated with information security. The focus is on incident responses, disaster recovery, and business continuity. The chapter explains risks, strategies, communication, and organization policies for managing security and risk for an organization. Students will utilize tools to identify and manage security risks. From a knowledge standpoint, we will review risk tolerance, processes for managing risks, impact of risks, security incidents, review disaster recovery planning processes, risk responses, tools for tracking risks, and risk metrics. As part of skills for the chapter, we will review the process for creating incident response plans, disaster recovery plans, risk models, and risk assessments. The application component of this chapter will focus on developing a risk assessment, security awareness plan, and impact assessment plan.

Knowledge Sections

Security and Risk Management and Planning

The process of security and risk management and planning are extremely important when dealing with digital transactions. It is important for organizations to identify risks that may be included in future forensic investigations. The approach for planning the risk activities is important and should be budgeted by the organization. Once the risks are identified, then the organization needs to establish a management process for the security concerns. Additionally, these concerns should be reviewed daily, weekly, or monthly depending on the organizational culture and risk tolerance. This is usually where the problem exists because risks are identified, but not managed with a consistent frequency. The approach should be to continually identify and control the risks. Security risks can derive from web, mobile, network, database, and local activity. These risks can potentially lead to digital investigations. Therefore, the process of defining risk needs to include an array of individuals to identify potential issues. The following provides a short list of potential resources we can use to identify risks for future forensic activities.

Security Risk and Forensics Resources
- Certified Public Accountants
- Attorneys
- Information Technology Professionals
- Digital Forensic Investigators
- Senior Staff
- Compliance and HR

The key is to have a plan and proper communication when these potential threats arise. We see threats arise from cyber incidents, data loss, compliance issues, social engineering attacks, denial of service, and impersonations. There is also a focus on the digital risk of system configuration issues, unauthorized access, and human threats. We need to understand the conflicts of interest, money laundering, intellectual property, corruption allegations, and fraud. Undoubtedly, we can usually associate risk with the assets of the organization. For instance, we know many transactional processing systems contain large amounts of customer and sensitive information. Therefore, we should identify the potential risks associated with the environment.

There are common steps we need to do when identifying the risks for our plans. We first need to determine how we plan to store the incidents, who has access to these incidents, how we protect the data, and the secure procedures associated with our infrastructure and networks. As shown below, there are several phases an organization can work through when conducting forensic investigations.

Common Forensics for Risk
- Discovery
- Digital Forensics
- Examination Solutions
- Forensics Analysis

As organizations have found, internal employees sometimes are the greatest concern for the organization. Therefore, the process of managing security and risk starts with developing clear and sufficient security policies and communicating these procedures and expectations to all employees.

Risk Management Approaches

Although we document and communicate our potential risk, there will be times where a particular issue does occur. In this case, we have to act on the risk according to our risk response document. As mentioned, our risk response document will identify the expected response and responsible parties if the risk does occur. In this case, we should have identified whether we want to accept the risks, mitigate, or transfer the risk. If we accept the risks, then we take the necessary actions to resolve the particular issue. In the case of mitigation, we can review multiple options to address the issue. However, there are many times when we use outside consultants or insurance policies. The following shows potential strategies we can use when dealing with risk.

Risk Management Approaches
- Mitigation
- Transference
- Acceptance
- Avoidance

The decision to select an approach needs to align with the overall organizational views and culture depending on the probability and impact of the risk. At this time, it is important to get the appropriate personnel involved if the risk can ultimately damage the organization's operations, reputation, or security.

Risk Tolerance

A key measurement for any organization is determining the overall risk tolerance for the culture and environment. Some organizations take a proactive approach opposed to a reactive approach in regards to risk. We can understand the organizational risk by surveying employees and understanding the overall culture. A few things you want to consider is how long the information system has been in existence and what are the mitigating factors if potential issues arise. These issues can stem from fraudulent activity, internal employee access of data, or external unauthorized activities. Organizations can conduct risk assessments on the environment to determine the potential risk and probability of the occurrences. This takes a number of reviews by a team to outline and document the potential risks. Additionally, it needs an incremental approach to realize the potential events that can have negative outcomes.

Risk Policies

Many of the allegations or instances occur when policies are not clearly outlined and documented within the organization. Therefore, organizations struggle with instances of security and risk when there is insufficient security policy, communication, and procedures. Another factor associated with this is the limited resources dedicated to develop and enforce such policies. Table 4.1 explains process of identifying risk, establishing governance, security networks, and developing a response to have clear guidance on the organization approach and stance regarding risks and protections.

Organizations can also participate in self-assessments to identify potential risk within the organization. This can be in the form of a questionnaire to collect data and information regarding potential concerns. We commonly see self-assessments conducted on the infrastructure before software development or implementations. However, it is important to provide this type of assessment holistically to help with potential security issues that will require an assortment of forensics or investigations. This activity also provides practice by examining files and processes for identifying, collecting, reviewing, storing, and reporting the potential activities. The items below are sample questions for a risk assessment survey.

Table 4.1 Risk plan

Action	Description
Identify Risk	Using diagrams, meetings, observations, and collaboration to identify potential risk associated with the organizational environment.
Establish Governance	Building a team within the organization to establish the security governments to help prevent future problems. This season include an assortment of employees from lower to higher levels staff.
Secure the Network	Reviewing and securing the current state of the network to include both physical and electronic access.
Develop Risk Response	Developing and documenting potential paths to address a particular issue if it were to occur.

Potential Risk Self-Assessment Survey

1. How are potential risks identified?
2. Where are the risks collected and stored?
3. How often is input gathered to update or remove risks?
4. What is the formal process or procedure when a potential risk is triggered?
5. Who is part of the risk assessment team?

As discussed, it is extremely important to properly plan with any security or risk initiative implemented within the organization. Therefore, we have to have overall acceptance and a formal budget to build plans such as the security, risk management, incident response, disaster recovery, and awareness. These initiatives will have less success if there is not a dedicated budget and resources allocated to the overall objectives. It is also critically important for top management to support these activities or they will not be adopted and utilized by the departments.

It is vital for organizations to document the types of risk and concerns for the entity. We find the procedures and descriptions in the Risk Management Plan. A sub component of the Risk Management Plan is a Risk Register. It is easy and important to utilize a document like the Risk Register. This document allows the organization to identify the risk & responsible individuals to the potential activities.

Luckily, there are many tools we can use to document and identify risks. A Risk Register is usually the basic form that can be developed individually or as a group. A collection of responses from the organization is a preferred method to ensure others are involved with the identifications. Common sections of this tool are document items, risks, triggers, responsibilities, consequences, probabilities, mitigations, and statuses. As shown in Table 4.2, this type of tool is simplistic, but provides a great matrix on the potential risk of the organization.

There are many Risk Management Tracking software options such as Intelex, File Handler, CammRisk, Compliance Manager, and Ideagen Risk Management along with many other options. These software options provide risk tracking, workflow rules, scoring, business intelligence tools, risks evaluation, and real-time dash boards. The risk metrics for these systems can come in forms of probabilities, and sample deviations,

Table 4.2 Risk register

Risk	Trigger Event	Responsibility	Consequence	Probability	Mitigation	Status

Risk Assessment and Model

Organizations build assessment and models to review the likelihood and risk levels. This takes a dedicated effort to evaluate the internal environment, external influences, technical constraints, and unknown events. These models can be developed using qualitative or quantitative approaches. Additionally, models can be broken up into different categories such as quality, or urgency assessments. Qualitative approaches can come in the form of face to face interviews, observations, or discussions while quantitative approaches may involve sophisticated software such as Barbecana, Palisade, or Vose. However, it is important to state some of the risk assessment and modeling software can have significate license cost, but identifying the risks will outweigh the possibility of not knowing the probability of high impact occurrences. We can conduct sensitivity analysis, monetary value analysis, Monte Carlo analysis, and decision trees using sophisticated risk analysis software. These are usually standard plug-ins or additional features depending on the platform and license that is purchased. The following is a list of sections that can be included in both a qualitative and quantitative assessment.

Inputs for Risk Assessment
- Type of risk event
- Risk exposure to organization
- Probability of the risk
- Potential costs if risk occurs
- High impact risks

Organizations have the option to utilize several forms of risk assessment. There are many best practices, but the key is to have well define approaches established in the organization to ensure risk are being assessed properly throughout the departments.

Incident Response Planning

The organization needs to document the process for responses if there is an incident. Incident response occur after an attack or risks identified in the risk planning is active. The organization should create a document with steps to complete once the incident is effective. The SANS Institute (2012) outlined the incident checklist as preparation, identification, containment, eradication, recovery, and lessons learned. This type of checklist provides information on communication to members regarding the security policies, where the incident occurred, containment solutions, backup and alternatives, approaches for recovery and documenting how the organization responded to the particular incident. The following shows a methodology for creating an incident response plan.

Incident Response Phases

1. Preparation
2. Identification
3. Containment
4. Eradication
5. Recovery
6. Documentation

As with any initiative, the organization will need to assign resources to incident response plan activities. It is important for the organization to notify law enforcement depending on the situation. However, there are procedures that you follow in regards to the jurisdiction. This can become complicated when we identify issues that are associated outside the country, state, or county lines. The following are resources that can be included as a team for these responses.

Incident Response Team Resources

- Incident Response Team
- Internet Service Providers
- Incident Reporters
- Law Enforcement Agencies
- Software Vendors
- Customers or Clients

As discussed in this chapter, it is great to have resources to form the team, but management needs to fully support the initiative to ensure the incident response team functions properly. Lack of resources on the team will only decrease the overall quality and value of the entity.

Disaster Recovery

Every organization should have in a place a Disaster Recovery Plan (DRP) in the case of an event that significantly impacts the organization business operations. We commonly think of natural disasters as the motivator for building a DRP. A significant disaster can be detrimental for long-term existence depending on the severity of the event. The items below outline common sections included in a DRP plan. However, the detail and sections vary depending on the market and type of business.

DRP Sections

- Plan Objectives
- Personnel
- Disaster response actions
- Identification of recovery site

- Inventory of hardware and software
  - Identification of assets established or taken to recovery site
- Summary of insurance coverage and liabilities
- Customer response plan

Cyber Security Awareness Plan

With a cyber-security plan, we can look at ways for prevention, resolution, and restitution. In these types of plans, we want to establish the policies for using the Internet, social media, and Internet browsing. Clear guidelines on the expected usage will improve the employee's overall knowledge of what is acceptable. However, this does not automatically translate to removing risks with the digital transactions. In such a plan, we can address the network usage, cloud-based services, password policies, handling of sensitive information, along with other technology security practices. It is advisable to build plans like this for the organization since it requires an investigation of the current environment and promotes communication in regards to cyber security issues.

Cyber Security Plan

1. Identify Roles and Responsibilities
2. Identify Key Terms
3. Data Security
4. Network and Cloud Security
5. Wireless Security
6. Web Security
7. Email Security
8. Mobile or BYOD Security
9. Physical Location Security
10. Employee Security

We tend to see buy in and acceptance when budgets are created for risk management and control. Risk continue to evolve and become more complex so it is important for organizations to send their IT personnel to training for the latest threats. Many organizations will state that their digital security is extremely important, but may not allocate enough funds to properly manage the risk along with resources needed for the risk identification or digital forensics.

Security Risk and Forensics Resources

- Certified Public Accountants
- Attorneys

- Information Technology Professionals
- Digital Forensic Investigators
- Senior Staff
- Compliance and HR

The key is to have a plan and proper communication when these potential threats arise. We see threats arise from cyber incidents, data loss, compliance issues, social engineering attacks, denial of service, and impersonations. There is also a focus on the digital risk of system configuration issues, unauthorized access, and human threats. We need to understand the conflicts of interest, money laundering, intellectual property, corruption allegations, and fraud. Undoubtedly, we can usually associate risk with the assets of the organization. For instance, we know many transactional processing systems contain large amounts of customer and sensitive information. Therefore, we should identify the potential risks associated with the environment.

There are common steps we need to do when identifying the risks for our plans. We first need to determine how we plan to store the incidents, who has access to these incidents, how we protect the data, and the secure procedures associated with our infrastructure and networks. As shown below, there are several phases an organization can work through when conducting forensic investigations.

Common Forensics for Risk

- Discovery
- Digital Forensics
- Examination Solutions
- Forensics Analysis

As organizations have found, internal employees sometimes are the greatest concern for the organization. Therefore, the process of managing security and risk starts with developing clear and sufficient security policies and communicating these procedures and expectations to all employees.

Risk Management Approaches

Although we document and communicate our potential risk, there will be times where a particular issue does occur. In this case, we have to act on the risk according to our risk response document. As mentioned, our risk response document will identify the expected response and responsible parties if the risk does occur. In this case, we should have identified whether we want to accept the risks, mitigate, or transfer the risk. If we accept the risks, then we take the necessary actions to resolve the particular issue. In the case of mitigation, we can review multiple options to address the issue. However, there are many times when we use outside consultants or insurance policies. The following shows potential strategies we can use when dealing with risk.

- Mitigation
- Transference
- Acceptance
- Avoidance

Risk Tolerance

A key measurement for any organization is determining the overall risk tolerance for the culture and environment. Some organizations take a proactive approach opposed to a reactive approach in regards to risk. We can understand the organizational risk by surveying employees and understanding the overall culture. A few things you want to consider is how long the information system has been in existence and what are the mitigating factors if potential issues arise. These issues can stem from fraudulent activity, internal employee access of data, or external unauthorized activities. Organizations can conduct risk assessments on the environment to determine the potential risk and probability of the occurrences. This takes a number of reviews by a team to outline and document the potential risks. Additionally, it needs an incremental approach to realize the potential events that can have negative outcomes.

Incident Response Planning

The organization needs to document the process for responses if there is an incident. Incident response occurs after an attack or risk is identified in the risk planning is active. The organization should create a document with steps to complete once the incident is effective. The SANS Institute (2012) outlined the incident checklist as preparation, identification, containment, eradication, recovery, and lessons learned. This type of checklist will provide information on communication to members regarding the security policies, where the incident occurred, containment solutions, backup and alternative approaches for recovery and documenting how the organization responded to the particular incident. The following shows a methodology for creating an incident response plan.

Incident Response Phases

1. Preparation
2. Identification
3. Containment

4. Eradication
5. Recovery
6. Documentation

As with any initiative, the organization will need to assign resources to incident response plan activities. It is important for the organization to notify law enforcement depending on the situation. However, there are procedures that we follow in regards to the jurisdiction. This can become complicated when we identify issues that are associated outside the country, state, or county lines. The following are resources that can be included in a team for these responses.

Incident Response Team Resources

- Incident Response Team
- Internet Service Providers
- Incident Reporters
- Law Enforcement Agencies
- Software Vendors
- Customers or Clients

It is also a good practice to send out a risk assessment survey to gain more insights on the potential risks and possible security vulnerabilities. These surveys can be sent electronically or conducted in a group or meeting setting. Although this is considered a best practice, there needs to be a group of individuals dedicated to collecting and reviewing the responses. The following provides a few sample questions that can be used for the risk survey.

Potential Risk Self-Assessment Survey

1. How are potential risks identified?
2. Where are the risks collected and stored?
3. How often is input gathered to update or remove risks?
4. What is the formal process or procedure when a potential risk is triggered?
5. Who is part of the risk assessment team?

As discussed, it is extremely important to properly plan with any security or risk initiative conducted within the organization. Therefore, we have to have overall acceptance and a formal budget to build plans such as the security, risk management, incident response, disaster recovery, and awareness. These initiatives will have less success if there is not a dedicated budget and resources allocated to the overall objectives. It is also critically important for top management to support these activities or they will not be adopted and utilized by the departments.

It is vital for organizations to document the types of risk and concerns for the entity. We find the procedures and descriptions in the Risk Management Plan. A sub component of the Risk Management Plan is a Risk Register. It is easy and important

Table 4.3 Risk register

Risk	Trigger Event	Responsibility	Consequence	Probability	Mitigation	Status

to utilize a document like the Risk Register. This document allows the organization to identify the risk & responsible individuals to the potential activities.

Luckily, there are many tools we can use to document and identify risks. A Risk Register is usually the basic form that can be developed individually or as a group. A collection of responses from the organization is a preferred method to ensure others are involved with identifications. Common sections of this tool are document items, risks, triggers, responsibilities, consequences, probabilities, mitigations, and status. As shown in Table 4.3, this type of tool is simplistic, but provides a great matrix on the potential risk of the organization.

There are many Risk Management Tracking software options such as Intelex, File Handler, CammRisk, Compliance Manager, and Ideagen Risk Management along with many other options. These software options provide risk tracking, work-flow rules, scoring, business intelligence tools, risks evaluation, and real-time dash boards. The risk metrics for these systems can come in forms of probabilities, and sample deviations.

Disaster Recovery

Every organization should have in a place a Disaster Recovery Plan (DRP) in the case of an event that significantly impacts the organization business operations. We commonly think of natural disasters as the motivator for building a DRP. A significant disaster can be detrimental for long-term existence depending on the severity of the event. The items below outline common sections included in a DRP plan. However, the detail and sections vary depending on the market and type of business.

DRP Sections

- Plan objectives
- Personnel involved
- Disaster response actions

- Identification of recovery site
- Inventory of hardware and software

    – Identification of assets established or taken to recovery site

- Summary of insurance coverage and liabilities
- Customer response plan

It is unfortunate that many small and medium size businesses do not have established DRP plans in place. Many organizations are stretched thin with their resources so developing disaster plans may be an afterthought. However, the potential loss of data and business operations should be taken into account. At the minimum, the risk awareness and plan should have content and communication strategies related to disasters.

Cyber Security Awareness Plan

With a cyber-security plan, we can look at ways for prevention, resolution, and restitution. In these types of plans, we want to establish the policies for using the Internet, social media, and Internet browsing. Clear guidelines on the expected usage will improve the employee's overall knowledge of what is acceptable. However, this does not automatically translate to removing risks with the digital transactions. In such a plan, we can address the network usage, cloud-based services, password policies, handling of sensitive information, along with other technology security practices. It is advisable to build plans like this for the organization since it requires an investigation of the current environment and promotes communication in regards to cyber security issues.

Cyber Security Awareness Components

1. Identify Roles and Responsibilities
2. Identify Key Terms
3. Data Security
4. Network and Cloud Security
5. Wireless Security
6. Web Security
7. Email Security
8. Mobile or BYOD Security
9. Physical Location Security
10. Employee Security

We tend to see buy in and acceptance when budgets are created for risk management and control. Risk continues to evolve and become more complex so it is important for organizations to send their IT personnel to training for the latest threats. Many organizations will state that their digital security is extremely important, but may not allocate enough funds to properly manage the risk along with resources needed for the risk identification or digital forensics.

Skill's Section

Discuss Tools for Tracking Risks

Activities/Indicators

- Research the top vendor software for managing and tracking risks.
- Develop a five-page paper discussing the backgrounds, current use, and future treads of the tracking software.
- Ensure to cite appropriately using the Author, YYYY citations.

Identify Risk Metrics

Activities/Indicators

- Create a document illustrating possible metrics that can be used to evaluate risk and control limits.
- Discuss the actions to take if a particular metric extends past a control limit.
- Ensure to cite appropriately using the Author, YYYY citations.

Resources

http://csrc.nist.gov/cyberframework/rfi_comments/040913_safegov_mwm_part_2.pdf
https://www.sans.org/reading-room/whitepapers/auditing/introduction-information-system-risk-management-1204

Identify Types of Security Incident Responses

Activities/Indicators

- Read the "The Incident Handlers Handbook".
- Outline the phases and activities associated with an incident response plan.
- Ensure to provide proper Author, YYYY citations with outside content.

Resources

https://www.sans.org/reading-room/whitepapers/incident/incident-handlers-hand-book-33,901 SANS: Preparation, Identification, Containment, Eradication, Recovery, Lessons Learned

Discuss why Disaster Recovery Planning Is Necessary

Activities/Indicators

- In a five-page paper, explain how the use of Cloud Computing can assist with disaster recovery planning.

- What are the potential cost associated with using a cloud environment for disaster recovery?
- What are the dangers?
- What industry would this strategy work well with?
- Should this be a public, private, or hybrid cloud approach?
- Ensure to provide proper Author, YYYY citations with outside content.

Resources
https://www.ready.gov/business/implementation/IT
https://www.nist.gov/sites/default/files/documents/itl/cloud/NIST_SP-500-291_
Version-2_2013_June18_FINAL.pdf

Evaluate Development Process for Risk Management

Activities/Indicators
Scenario: You work in IT as a risk analyst for a financial institution. You notice the department has guidelines and procedures associated with digital forensics. However, there is not much information on types of risk associated with the activities.

- In a MS Word document, identify the potential issues and risks that can arise from a forensics investigation.
- Explain any liabilities the organization may face once an investigation has initiated.
- Ensure to cite appropriately using the Author, YYYY citations.

Create an Incident Response Plan

Activities/Indicators

Scenario: You work for a retail company and you have been asked to develop and incident response plan. The retail company sells high end bicycles to consumers both on-site and online. The focus of the incident response plan should be on the availability of the Internet and network accessibility for the location.

- Develop an Incident Response Plan for the medium size business.
- Identify the phases and resources associate with the plan for an ecommerce site.
- Ensure to provide proper Author, YYYY citations with outside content.

Resources
https://www.sans.org/reading-room/whitepapers/incident/incident-handlers-handbook-33901

Develop a Disaster Recovery Plan

Activities/Indicators

Scenario: You work for a medium size engineering firm that houses sensitive government contracts and building specification. The drawings, content, and contracts are secured in the network, but there is not a formal DRP in place. Your responsibility is to provide a recommendation and alternative recommendation to ensure the organization has a DRP in place.

- Develop a Disaster Recovery Plan for the engineering firm.
- Include the appropriate sections with justifications.
- Ensure to provide proper Author, YYYY citations with outside content.

Resources
https://www.ready.gov/business/implementation/IT
https://www.nist.gov/sites/default/files/documents/itl/cloud/NIST_SP-500-291_Version-2_2013_June18_FINAL.pdf

Applications' Section

Develop a Risk Assessment or Model

Activities/Indicators

Scenario: You work in IT for a successful in Oil and Gas organization. You have been charged to create a Risk Assessment or Model for your organization.

- In Microsoft Visio or another diagraming software, diagram the process of system characteristics, threat identification, vulnerability identifications, controls, probability, impacts, recommendations, and documentation.
- In MS Word, explain how the model can be used with both qualitative and quantitative analysis.
- Ensure to cite appropriately using the Author, YYYY citations.

Resources
http://csrc.nist.gov/publications/nistpubs/800–30/sp800–30.pdf

Create a Cyber Security Plan

Activities/Indicators

Scenario: Create a Cyber Security Plan for a small medical office. Ensure to include the sections identified in the Cyber Security Plan.

- In a seven-page report, outline the sections along with sufficient content for each potential security area.
- Discuss how the cyber security connects to the risk management plan.
- Include a Reference page for research.
- Include a Glossary for unique terms.
- Ensure to provide proper Author, YYYY citations with outside content.

Resources

https://www.dhs.gov/sites/default/files/publications/FCC%20Cybersecurity%20Planning%20Guide_1.pdf

Questions

- Discuss the components of Disaster Recovery Plan. Identify the major areas of focus.
- Outline the components of a Business Continuity Plan.
- Identify and discuss the sections for Risk Metrics.
- Identify and discuss the Risk Management Plan sections.
- Explain the types of risks associated with an ecommerce project.
- Outline the Disaster Recovery process for a small business.
- Develop a Business Continuity Plan for real estate broker with five employees.
- How can departments, organizations, and outsourced vendors have shared risk with an information system?
- Outline the Risk Management sections.
- Explain the IS Security Risk concerns with healthcare data.
- Explain the process of assigning incident response duties to employees.

Chapter 5
Encryption and Information Protection/Integrity and Concealment Methods: Lesson Plans

Competency: Learn major aspects and concepts of cryptography
Activities/indicators
• Study reading material provided by instructor related to major aspects of **cryptography**.
• Complete successfully an assessment provided by instructor related to competency content.
• For mastering levels, more than 80% of assessment grades should be earned in no more than three trials.
• Assessment questions can be pulled from the end of the chapter questions or any relevant material.
Competency: Learn major applications of cryptography
Activities/indicators
• Study reading material provided by instructor related to major aspects of **applications of cryptography**.
• Complete successfully an assessment provided by instructor related to competency content.
• For mastering levels, more than 80% of assessment grades should be earned in no more than three trials.
• Assessment questions can be pulled from the end of the chapter questions or any relevant material.
Competency: Learn major aspects and concepts of information integrity (i.e. hashing, etc.)
Activities/indicators
• Study reading material provided by instructor related to major aspects **and concepts of information integrity (i.e. hashing, etc.)**.
• Complete successfully an assessment provided by instructor related to competency content.
• For mastering levels, more than 80% of assessment grades should be earned in no more than three trials.
• Assessment questions can be pulled from the end of the chapter questions or any relevant material.
Competency: Learn major applications of information integrity (i.e. hashing, etc.).
Activities/indicators
• Study reading material provided by instructor related to major applications **of information integrity (i.e. hashing, etc.)**.
• Complete successfully an assessment provided by instructor related to competency content.
• For mastering levels, more than 80% of assessment grades should be earned in no more than three trials.
• Assessment questions can be pulled from the end of the chapter questions or any relevant material.

(continued)

© Springer International Publishing AG 2018
I. Alsmadi et al., *Practical Information Security*,
https://doi.org/10.1007/978-3-319-72119-4_5

Competency: Learn major aspects and concepts of information concealment (i.e. steganography, watermarking, etc.)

Activities/indicators
- Study reading material provided by instructor related to major aspects **and concepts of information concealment (i.e. steganography, watermarking, etc.)**
- Complete successfully an assessment provided by instructor related to competency content.
- For mastering levels, more than 80% of assessment grades should be earned in no more than three trials.
- Assessment questions can be pulled from the end of the chapter questions or any relevant material.

Competency: Learn major applications of information concealment (i.e. steganography, watermarking, etc.)

Activities/indicators
- Study reading material provided by instructor related to major applications **of information concealment (i.e. steganography, watermarking, etc.)**
- Complete successfully an assessment provided by instructor related to competency content.
- For mastering levels, more than 80% of assessment grades should be earned in no more than three trials.
- Assessment questions can be pulled from the end of the chapter questions or any relevant material.

Competency: Use and evaluate encryption algorithms.

Activities/indicators
- Use one or more open source or free tools that allow users to create/modify encryption algorithms.
- For mastery, student is expected to try more than one encryption algorithm.
- For a mastering level, student should show what tool they selected, how they installed the tool and/or present a video to demonstrate the different commands they have used.

Competency: Use and evaluate encryption applications.

Activities/indicators
- Use one or more open source or free tools that allow users to create/modify encryption tool/application.
- For mastery, student is expected to try more than one encryption tool/application.
- For a mastering level, student should show what tool they selected, how they installed the tool and/or present a video to demonstrate the different commands they have used.

Competency: Use and evaluate different hashing/steganography algorithms

Activities/indicators
- Use one or more open source or free tools that allow users to create/modify hashing/ steganography algorithms.
- For mastery, student is expected to try more than one encryption tool/application.
- For a mastering level, student should show what tool they selected, how they installed the tool and/or present a video to demonstrate the different commands they have used.

Competency: Use and evaluate different hashing/steganography applications.

Activities/indicators
- Use one or more open source or free tools that allow users to create/modify hashing / steganography applications.
- For mastery, student is expected to try more than one encryption tool/application.
- For a mastering level, student should show what tool they selected, how they installed the tool and/or present a video to demonstrate the different commands they have used.

Competency: Select an information system/website, etc. and evaluate it from information reliability, and validity perspectives

(continued)

Activities/indicators

- Students should select a case study from their work environment or any simulated one to evaluate information reliability and validity.
- Students should present a critique and evaluation based on a base-line or a national standard.
- Students should also propose on what policies should be changed/updated or added for existing business policies to accommodate most recent standards.

Competency: Select an information system/website, etc. and evaluate it from information privacy and protection perspectives

Activities/indicators

- Students should select a case study from their work environment or any simulated one to evaluate information privacy and protection.
- Students should present a critique and evaluation based on a base-line or a national standard.
- Students should also propose on what policies should be changed/updated or added for existing business policies to accommodate most recent standards.

Overview

Information protection is a key goal to most of information security controls. Without information hiding and protection mechanisms the whole e-commerce and remote communication mechanisms will not be possible. As users in the Internet communicate without seeing each other, mechanisms should exist to validate the identity of each party of the communication activity to the second party. Their communication and all information exchanged between them should be also protected so that no third party can possible see or expose such data or information.

In this chapter, we will focus on three important activities related to information processing: Information hiding or protection through encryption and information integrity verification through hashing and information concealment through steganography and watermarking.

Knowledge Sections

Information Hiding and Protection: Cryptography or Encryption

The basic concepts on encryption or information hiding are very old that go back to very early ages of human history. In its basic definition, encryption or information hiding includes the following basic elements (Fig. 5.1):

1. Two points of communication: In most cases, an encryption communication or text will occur between two and only two communication points (e.g. humans, devices, websites, etc.). Those two don't have to be symmetric of

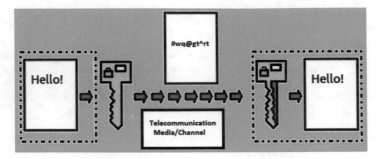

Fig. 5.1 Basic encryption reference model

course. For example, this communication can be between a human and a website. In some cases, communication points can be one-to-many, many-to-one or many-to-many. However, in most cases, those "many" are symmetric and represent one entity. For basic terminologies, lets define those two communication points as "sender" and receiver" and we will call the information or text to send as "message".

2. A communication channel: Through history, people used to send encrypted messages to remote or distant receivers. The assumption here is that you don't have any control in the communication channels and many "strangers" or "intruders" may try to intercept this message, which is usually very sensitive or "classified". As the process of "encryption" is usually expensive and will require special resources, it will not be used to send a message which is "public" in nature as this will not justify why to invest resources in trying to hide such information while there is no serious consequence of strangers saw/view or intercept such information or message.

3. Encryption process or method: A message that will be sent through a public communication channel "such as the Internet", should be hidden or encrypted in such a form that no one in this open or public channel can "understand it" although they can "see" or "view" it.

We can view human languages as a form of encryption. For example, if two persons next to you are speaking a language that you don't understand, you will hear their voices, but you will not understand what they are talking about or their message. The only problem with this definition is that a language is usually known in the world by more than two (much more) and hence if you want such message to be private, you can't send it through the public Internet where there will be many human who can understand the language or your encryption method or algorithm.

Based on this general model or requirements for information hiding or encryption, humans through history used different forms and techniques to convey and protect messages send through the air. The air was the main media of telecommunication, not transportation, until the recent invention of phones, telegraphy, etc.

For example, humans in history used smoke, bills, animals, drums, etc. to convey messages to distant humans. Unless you know previously such convention, as a human you can see/hear, etc. smoke, drums, etc. but not necessary understand the message behind them.

In the telecommunication ages, encryption used largely by military, governments, and related agencies to carry confidential or top secret messages.

Methods to Break Encrypted Text

- Simple guessing: This can work for small size messages or text.
- Through interception: Getting either the key or the original text by accessing the communication between the sender and recipient or gaining access their computer.
- Brute Force: Until a correct match is made all the random keys are tried using computing powers.

Ciphering or Encryption Algorithms

Most of the components in the encryption system or reference architecture can be considered as generic in their general functionalities. The exception to this is the ciphering or encryption algorithms. Those are the methods or techniques that decide how original text will be converted to encrypted text before sending it through the telecommunication media and how it will be decrypted back at the receiver side or destination.

There are different classifications for encryption algorithms. In one popular classification, encryption algorithms can be broadly classified into different types:

Key-based Encryption techniques
Those can be broadly classified into: Symmetric and Asymmetric algorithms:

Symmetric Encryption Algorithms

The term symmetric indicates that the same keys are used for encryption and decryption. Those can be further classified into: Block and Stream ciphers:

Block Ciphers

Another criterion that characterizes the encryption algorithm is the block size. For example, DES block size is 64bits. This means that the size of encryption key is less than or equal to 64 bits. Specifically, 56bits are used for the key and the rest are used for parity check or error correction. Examples of block cipher known encryption algorithms are: DES, AES, Blowfish, etc.

- DES: One example of this type is: Digital Encryption Standard (DES). Sender and receiver are using the "same" key for the encryption and the decryption process. Early implementations of DES started by IBM (Lucifer algorithm) in 1973.

DES can be also classified under the classification (product cipher or product block cipher) in which two or more simple ciphering algorithms (such as substitution or transposition/permutation ciphers) are combined, through several iterations or rounds (e.g. 16) to produce a stronger algorithm.

In DES, block size is 64-bits. Triple DES, a new DES encryption algorithm, has a 192-bit key (but only uses 168-bits of them). The number of bits in the key also decides how strong is the encryption and how long it takes to break that key. While it will take millions of years to break a 64-bit key encryption if the process is executed in one serial thread, the process can be significantly accelerated using multithreading or parallel programming and processing. This block size decides also how much data can be encrypted or transformed from plain to encrypted text. DES uses Feistel structure in which, for each round, one-half of the block is unchanged and the other half goes through a transformation that depends on the S-boxes and the round key.

Figure 5.2 shows an example of a simple 64-Bit DES encrypted text using the online encryption website (https://www.tools4noobs.com/online_tools/encrypt/).

From Fig. 5.2, we can observe the following issues that decide DES encryption output:

- DES uses two types of encoding: 64Bit and Hexa.
- DES uses different modes such as: CBC, CFB, CTR, ECB, etc. Those can also decide the nature of the encrypted text. DES modes are:

  - ECB: Electronic Code Book: This is considered a simple deterministic encryption scheme. In ECB, each block is encrypted independently.

Fig. 5.2 An example of 64-Bit DES encryption

Fig. 5.3 CBC encryption mode

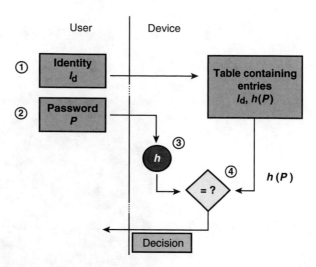

Additionally, for the same text, encryption will always be the same. Or in other words, if we repeated encrypting the same text several time, it will always create the same encrypted text. The deterministic nature of ECB made it as a desirable encryption scheme.

- CBC: Cipher Block Chaining: Unlike ECB, CBC is an un-deterministic encryption scheme. Figure 5.3 shows basic encryption CBC scheme. Initial Vector (IV) is selected randomly for each message. IV with Key is XORed with plain text to produce the cipher text. Repeating the same process for the same message can generate a different encrypted message. Changing IV results in different cipher-text for identical message.
- CFB: Cipher FeedBack: In CFB, the process is initially similar to CBC. However, the plain text is divided into several parts (sub-texts) and the key is XORed with sub-texts gradually rather than with the whole text one time (Fig. 5.4).

In CFB mode, encryption output for each stage does not only depend on initial plain text, key and IV, but also on previous encrypted text. This is why as IV is random, repeating the process for the same plain text will produce different cipher text every time.

- OFB: Output FeedBack: OFB is very similar to CFB. The only different is that the input to the next cycle is taken from the "block cipher encryption" instead of taking it from the ciphertext. Additionally, in CFB IV must be unpredictable unlike in OFB.
- CTR: Counter: In this mode, a counter that plays a similar role like the IV in CFB and CBC modes, is shared between encryption sender and receiver.

  • Users can select the "key" which can also decide the nature of the encrypted text.

Triple DES or 3DES appeared more recently as a replacement to classical DES where cipher algorithm key is applied 3 times to each data block. As a result, the

Fig. 5.4 CFB encryption
mode

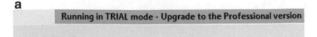

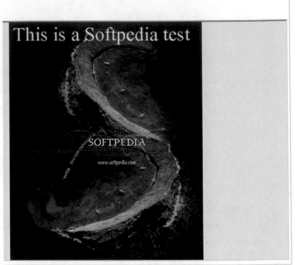

encryption key size will be 3 times larger than original size (i.e. from 56bits to 168bits. As most other current encryption algorithms are 128-bit, 3DES was an acceptable replacement to DES to keep the encryption alive and usable in the current market.

The standard took the number RFC 1851 and was approved in 1995. The encryption has different versions of implementation where the 3 keys can be independent or identical. 3DES works by taking 3 of 56-bit keys (e.g. K1, K2 and K3), and first encrypting with K1, decrypting next with K2 and encrypting third and last time with K3.

Some examples of applications of 3DES include some electronic payment systems, Microsoft in some of their applications, browsers such as Firefox, etc.

More recently, around the new century (1999), DES algorithm is phased out to more secure algorithms. National Institute of Standards and Technology (NIST) approved the Advanced Encryption Standard (AES) as a replacement for classical DES. NIST endorsed Triple DES as a temporary standard to be used until AES was finished. In comparison with 3DES, AES is faster.

- Advanced Encryption Standard (AES): A widely accepted encryption standard (currently). Another symmetric key block cipher. AES accepts keys of 128, 192 or 256 bits (In comparison to 64 in classical DES and 128 in 3DES). The most widely used one is the 128bit which is strong and long enough for any brute force technique to break it. In US, NIST decided to adopt AES in the US in 2001. The first advantage of AES over DES is the longer key which makes it much harder to break the encryption. The block size is larger which means also faster data transmission and larger possible message size to send. Another difference between DES and AES, is that DES is a bit-oriented cipher, AES is a byte-oriented cipher. This is another factor why AES implementation is faster than DES.

AES uses a combination of substitution and permutation instead of Feistel network used in DES. DES plaintext block is divided into two parts/halves before the main algorithm starts whereas, in AES the entire block is processed to obtain the cipher text. In the process, DES uses 16 rounds mostly based on XOR process. The first step of the cipher is to put the data into an array; after which the cipher transformations are repeated over a number of encryption rounds. The number of rounds is determined by the key length. AES uses 10 rounds for 128-bit keys,12 for 192-bit keys or 14 for 256-bit keys using the following processes: Subbytes, Shiftrows, Mix columns, and Addroundkeys processes (Fig. 5.5).

Here is a simple explanation of AES processes:

Fig. 5.5 AES encryption round

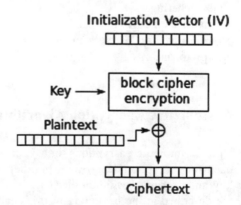

- Subbytes: This is a substitution process. It uses S-box by which it performs byte by byte substitution of the entire block or matrix.
- Shift Rows: Rows of the matrix are shifted.
- Mix Columns: Columns are of the matrix are shuffled from right to left.
- Add round keys: Here, the XOR of the current block and the expanded key is performed. Longer keys need more rounds to complete.

Except for the last round in each case, all other rounds are identical.The process of decryption of an AES ciphertext is similar to the encryption process in the reverse order.

BlowFish

BlowFish is also a symmetric key block cipher. Blowfish is a block cipher proposed by Bruce Schneier. Blowfish block size is 64bit just like DES. Blowfish uses a variable number of bits ranging from 16–448 bits and encrypts the data 16 times to make it hard for a hacker to break or decrypt it using brute force.

Stream Ciphers

In comparison with block cipher, stream cipher encrypts messages one bit or byte at a time rather than working on a block of data in block cipher. In one difference, stream cipher can suffer less from data noise as the size of data sent each time is much less.

Examples of stream cipher encryption algorithms include: RCA, RC4, SEAL, FISH, BMGI, etc.

The key-stream is generated serially from a random seed. In comparison with block cipher, stream cipher is less popular in terms of practical utilization or usage. One problem with stream cipher is the key size. In order to get a stream cipher key that is properly secure, it should be large, much larger than the typical key size for block ciphering.

Rivest Cipher 4 (RC4) is a popular stream cipher. RC4 is used in various protocols such as: Wireless encryptions: WEP and WPA as well as in TLS and SSL web protocols.

Asymmetric Encryption Algorithms

Unlike symmetric encryptions, in asymmetric encryptions, two different keys are used for encryption/decryption. In other words, message is encrypted with one key and decrypted with another key. In comparison with symmetric encryption, we described at the beginning of this chapter three methods to break encryption messages. While symmetric encryptions are subjected to all 3 types, asymmetric encryption can be only subjected to the last one (i.e. brute force).

Examples of asymmetric encryption algorithms include: RSA, ECDH, ECC, Diffie-Hellman, etc.

- **Rivest-Shamir-Adleman (RSA):**

RSA is probably the most popular asymmetric encryption. RSA uses a variable size encryption block and a variable size key. The algorithm is widely used in digital signatures and e-commerce. Two prime numbers are used to generate the public and private keys. The algorithm from those 3 authors first appeared in 1977 and published officially in 2000. However, it is believed that the idea of asymmetric pair-keys were discussed in the same period or before from other authors or researchers (e.g. Whitfield Diffie and Martin Hellman in 1976 and Clifford Cocks in 1973). RSA keys are typically 1024- or 2048-bits long.

In the algorithm, users can distribute their public keys to all their friends. They only need to keep their private keys confidential. Let's assume that sender A knows user B previously. In this case, A public key, pubA is known to user B. Similarly, B public key, pubB is known to user A. If A wants to send a message to B, he/she will encrypt the message with pubB. As a result, the only way to decrypt this message is by processing it with its pair (i.e. the private key for B, privateB). The only one who has this key, private, is the user B. This means that the only one who can retrieve or decrypt this message is user B. Figure 5.6 shows a simple example of RSA algorithm demo using RSAvisual tool. We can notice the followings based on RSA algorithm.

- The algorithm starts by picking two prime numbers: p and q (in practice those should be large numbers, we picked two small prime numbers for simplicity).
- The modulus (n) is calculated by simply multiplying p and q. This modulus number is used by both the public and private keys and provides the link between them. The modulus length, usually expressed in bits, is called the key length. The public key (that does not need to be a secret) consists of two parts:

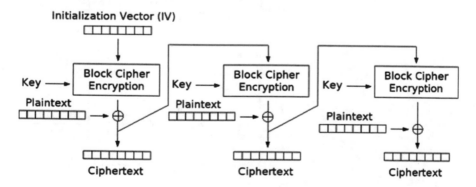

Fig. 5.6 RSA algorithm calculator

the modulus n, and a public exponent, e, which is normally set at 65537 (We used 17 for e for demo).

- The private key consists of the modulus n and the private exponent: d, which is calculated using the Extended Euclidean algorithm to find the multiplicative inverse with respect to the totient of n or $\varphi(n)$ (In some literature it is referred to as: r).
- M refers to the message size.
- The cipher text C is calculated as: Me mod n.
- P, q and e are input values selected randomly (within certain constraints). N, e and d are values used in the RSA output process, N and e for the public key, N and d for the private key.
- Attacker needs to factorize n to obtain p and q. Factorizing large numbers is difficult and slow. Brute Force attack is less efficient than factorizing.

Encryption Substitution Techniques

Substitution ciphers in general replace plaintext with cipher-text. This can be as simple as shifting the letters in the plaintext with a fixed number of shifts (as in Caesar cipher which is also called shift cipher). For example, if the offset value is s = 3 s = 3, then the plaintext "caesarcaesar" will be encoded as the ciphertext: "fdhvdufdhvdu".

Examples of such ciphers include: Caesar, Playfair, Hill, Rail Fence, etc. Figure 5.7 shows a simple example to demo Caesar substitution encryption. Caesar falls within a simple substitution type called: monoalphabetic substitution.

Fig. 5.7 A simple Caesar substitution encryption example

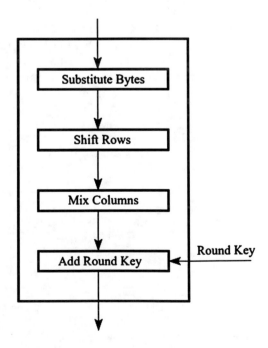

Fig. 5.8 Vigenere quadrat table

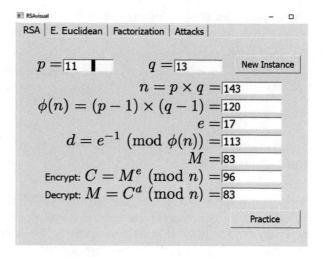

RSAvisual — ☐

RSA | E. Euclidean | Factorization | Attacks

$p = 11$ $q = 13$ New Instance

$n = p \times q = 143$

$\phi(n) = (p-1) \times (q-1) = 120$

$e = 17$

$d = e^{-1} \pmod{\phi(n)} = 113$

$M = 83$

Encrypt: $C = M^e \pmod{n} = 96$

Decrypt: $M = C^d \pmod{n} = 83$

Practice

To be or not to be, That is the question   Use key: 3 ▾

Encrypt / Decrypt

Output:

Wr eh ru qrw wr eh, Wkdw lv wkh txhvwlrq

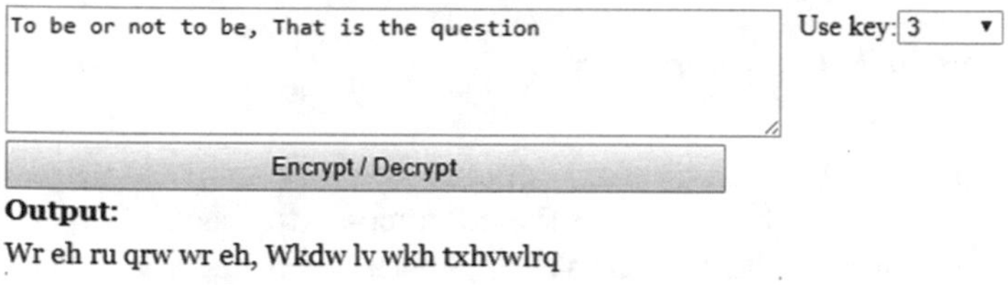

Fig. 5.9 A simple permutation example

In addition to monoalphabetic, polyalphabetic substitution can be used. Vigenere cipher is an example of polyalphabetic substitution. Vigenere uses two components, in addition to the plaintext, to create the encrypted text: A key of text, which can be of any size and Vigenere Quadrat table (Fig. 5.8).

The table matrix is used to find the encrypted text that matches plain text (column) and key (as the row). The advantage of this algorithm over Caesar cipher is the "key". The key is user defined and should be kept private. This makes it harder to brute force this approach in comparison with Caesar cipher. Of course this depends on the key selection. But as far as the key is more than one letter, it will be stronger than Caesar algorithm.

Substitution encryptions are old. They have been used for a long time and they are still in use for educational and demo purposes. However, in terms of practicality, they are considered as weak encryption methods that can be easily broke using brute force methods.

Similar to substitution, in permutation, plain text is shuffled or the places of letters/words in the plain text are shuffled to switch positions, randomly to produce the cipher text. Figure 5.9 shows a simple permutation example.

Cryptography Applications

The applications where cryptography is used are enormous and are continuously growing. We will select in this section some of the popular areas where encryption is used.

Encryption for Internet/Online Communications and e-Commerce

Encryption is about hiding data, but also the value of encryption is hidden and underestimated in many areas. For example, without encryption, the whole ecommerce wouldn't be possible as we see it today.

Encryption in E-Commerce

The major different between classical commerce and ecommerce is that unlike in traditional commerce, the parties of the ecommerce transaction do not have to see/know each other (in the classical sense). This is where encryption and its applications come in place. First, there is a need for a mechanism to uniquely identify users through online. Second, there is also a need to verify the identity of each user or party in real time. Third there is a need to ensure that communication/messages between the two parties of any online transaction are not visible to unintended audience or third-party intruders. Those lead to the major 4 goals of using encryption in e-commerce: Confidentiality, integrity, authenticity, and non-repudiation.

RSA encryption that we described earlier and related Public Key Infrastructure (PKI), or also called Public Key Encryption (PKE) is widely used in encrypted websites/online communications. Additionally, third parties called Certificate Authority (CA), or Trusted Third Party (TTP) such as: Verisign, RapidSSL, Thawte, etc. are used to verify the correct identity for online users.

The most common use of PKE for e-commerce involves the use of Digital Certificates issued by CAs or TTPs. Digital certificates are sent with messages to verify the identity of communication partners to each other. They can be in the form of an embedded file, created by special programs with hard-coded and encrypted information about users' machines. Many software applications now use this technique (digital certificates) to protect their licensed keys from being used on many machines. The first computer that has the valid key will create a digital certificate that will be sent to a special license server for activation. After this initial activation, the key will be uniquely tied with the machine, through the digital certificate.

Secure Sockets Layer (SSL) protocol allows for the transmission of encrypted data across the Internet by running above the TCP/It protocols. This is why some hyperlinks have the (s), (http) to indicate that communication in this page is encrypted using SSL or TLS. Many browsers added features to warn users to include their information if the website they are visiting is not secured.

SSL and its alternative Transport Layer Security (TLS) became the most popular secure communication or network protocols over the Internet. They are used through web browsing, email messages, e-commerce, Voice over IP VoIP, etc. In order to differential SSL HTTPS traffic from regular HTTP traffic, SSL HTTPS traffic typically uses port 443 (instead of 80 or 8080 for HTTP). SSL was first created through Netscape and their encryption expert: Taher Elgamal. Elgamal himself has under his name an asymmetric encryption algorithm that is integrated in some PKI applications. SSL went through 3 versions 1, 2 and 3. Currently HTTPS is not used by all websites. Some websites used it also partially, through the pages that require users to post back information to the website. SSL certificates are provided for websites at yearly costs with different types of subscriptions.

An alternative to SSL/TLS is to use Virtual Private Networks (VPNs). In comparison with SSL/TLS, VPN is recommended for two or more communication parties with permanent or frequent communication. A typical example will be bank branches, company different locations, etc.

Encryption in Email Communication

Encrypting email contents in transit and in servers is very important and vital. People nowadays trust that their email contents will only be reached to their intended receivers and will not be exposed whether in transition or in the email servers. Email systems use different types of encryption algorithms or methods including: TLS/SSL, PGP, S/MIME, GNU Privacy Guard (GnuPG), etc.

When you check your email using the web browser, the first thing that you need to check before login to your account is that the current page is secure (i.e. https, TLS/SSL), with a current and valid certificate. In client email applications such as MS outlook, you can enable SSL settings (Fig. 5.10).

Encryption in Browsers

Web browsers such as Google Chrome, Microsoft Internet explorer, Firefox, etc. are very important applications to enable users to reach different websites through the Internet. Those web browsers evolved significantly in the last few years. Some of the significant evolutions are related to security improvements in order to be able with the continuously growing security threats. Browsers represent client side applications in client-server architectures. Many protocols and services support the communication between server and client sides. Table 5.1 shows list of web browsers and their most recent supported ciphers (Ref., www.rapidssl.com, 2017). TLS/SSL

Vigenère-Quadrat

	Text																									
	A B C D E F G H I J K L M N O P Q R S T U V W X Y Z
1	A B C D E F G H I J K L M N O P Q R S T U V W X Y Z
2	B C D E F G H I J K L M N O P Q R S T U V W X Y Z A
3	C D E F G H I J K L M N O P Q R S T U V W X Y Z A B
4	D E F G H I J K L M N O P Q R S T U V W X Y Z A B C
5	E F G H I J K L M N O P Q R S T U V W X Y Z A B C D
6	F G H I J K L M N O P Q R S T U V W X Y Z A B C D E
7	G H I J K L M N O P Q R S T U V W X Y Z A B C D E F

Fig. 5.10 Enable SSL in MS outlook

Table 5.1 Web browsers and current supported ciphers

Web browser	Operating system	Ciphers
Internet Explorer 9	Windows Vista, 2008, Windows 7	AES-256-bit
Internet Explorer 9	Windows XP	RC4-128-bit
Internet Explorer 8	Windows 2008	AES-256-bit
Internet Explorer 8	Windows Vista	AES-256-bit
Internet Explorer 8	Windows XP	RC4-128-bit
Internet Explorer 7	Windows Vista	AES-128-bit
Internet Explorer 7	Windows XP	RC4-128-bit
Internet Explorer 6	Windows XP & 2003	RC4-128-bit
Firefox 4	Windows XP, Vista, 2008, Win 7	AES-256-bit
Firefox 3.6	Windows XP, Vista & Windows 2007	AES-256-bit
Firefox 3.01	Mac OSX 10.5.x	AES-256-bit
Safari v3.2.1	Windows XP	RC4-128-bit
Safari v3.2.1	Windows Vista	AES-128-bit
Safari v3.2.1	Mac OSX 10.5.x	AES-128-bit
Safari	IPhone v2.2	AES-128-bit
Chrome v1.0.154.43	Windows XP	RC4-128-bit
Chrome v1.0.154.43	Windows Vista	AES-128-bit
Opera v9.62	Windows XP	AES-256-bit
Opera v9.62	Windows Vista	AES-256-bit

encryption is used to secure or encrypt content through web pages. The different browsers include methods to alert uses if they are trying to post back information through web pages that do not support properly TLS/SSL. This is especially important as phishing through fake web pages is very popular. Attackers create fake web pages that look exactly like popular web pages/sites that users typically login to (e.g. emails, bank accounts, social networks, etc.). They sent their links to victim users through emails, social networks, etc. promoting users to login and post their credentials in those phishing links (Fig. 5.11).

Fig. 5.11 An example of a
phishing link

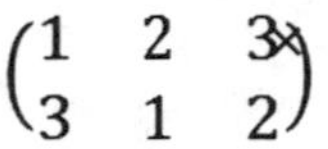

$$\begin{pmatrix} 1 & 2 & 3 \\ 3 & 1 & 2 \end{pmatrix}$$

Encryptions in Telecommunication/Networks

Network encryption at the network or data link layer is not visible to users and their applications. Internet Protocol Security (IPSec), a set of open Internet Engineering Task Force (IETF) standards is an example of encryption at this level. The goal is to create private communication over IP network. Encryption can occur at the packet level.

Virtual Private Networks VPNs are created through encrypting the traffic using virtual tunneling protocols. This enables remote connection or remote users to connect using the public Internet. Their traffic will be tunneled and encrypted independent from the rest of Internet traffic. An attacker who is trying to sniff or spy on the data will only see encrypted data which will ensure its confidentiality. This can also prevent unauthorized users from using the VPN.

VPN uses IPSec as a secure protocol for IPv6. TLS/SSL is also used to encrypt network traffic in some applications such as OpenVPN. Cisco uses another protocol: Datagram Transport Layer Security (DTLS) for UDP traffic. Secure Socket Tunneling Protocol (SSTP) is used by Microsoft for Point to Point Protocol PPP or communication.

Protocols such as SSH and OpenSSH are also used to secure network sockets or communications. Many applications that allow remote access, use SSH or similar protocols for secure connections.

Secure shell (SSH), is a secure protocol to provide safe administration of remote servers. SSH can be configured to use different symmetrical cipher systems, including AES, Blowfish, 3DES, etc.

SSH utilizes also asymmetric encryption in a few different instances. During the initial key exchange process used to set up the symmetrical encryption between connection parties (used to encrypt the session), asymmetrical encryption is used. In this stage, both parties produce temporary key pairs and exchange the public key in order to produce the shared secret that will be used for symmetrical encryption. If you are creating an SSH connection for a remote server, you can use some tools (e.g. Putty) to create your key pair and upload the public one to the remote server.

Encryption in Operating Systems and Disks

Disk encryption is used to encrypt every bit of data that is written to the disk or disk volume. This is used based on disk encryption software or hardware. Disk encryption is used to prevent unauthorized access to data storage. Disk encryption provides another layer of information protection (beyond the network, operating system, etc. access control). Disk encryption is also different from file-encryption that provides encryption per files as disk encryption software provides encryption for the whole disk, bit by bit, as one unit.

For file encryption, Microsoft used Encrypting File System (EFS) in their NTFS file system (Version 3.0). Files can be encrypted and protected from users who can have access to the desk or the operating system. EFS uses symmetric encryption algorithms which can be faster to encrypt and decrypt in comparison with asymmetric algorithms.

Followings are examples of popular disk encryption software:

- BitLocker for Windows: Microsoft started using BitLocker encryption in their operating systems from Windows Vista and later on. The system is using AES encryption with CBC or XTS (cipher-text stealing) mode.

There are two types of attacks to consider here: The attacker may try to know the information in the disk, and we want to prevent them from doing so. Second, the attacker may try to tamper such information, without being able to know it, and we want also to prevent them from doing so.

Using Elephant diffuser is Microsoft's proposed solution to CBC mode modification to allow it to prevent data tampering, the specific data tampering and not the random. The idea here is to "mix" the plaintext up before encrypting it, so that an attacker can't change specific data (called data meddling or malleability). Microsoft then removed Elephant diffuser from being the default cipher choice in Windows 8 and beyond (citing speed and standard compliance as reasons). The mode is kept for backward compatibility.

XTS is introduced in Microsoft operating systems in Windows 10. In comparison with CBC, there is no requirement for an initialization vector, IV for XTS. XTS tweak key can be derived from the block number. This tweak key can be sector address or a combination of the sector address and its index (Fig. 5.12).

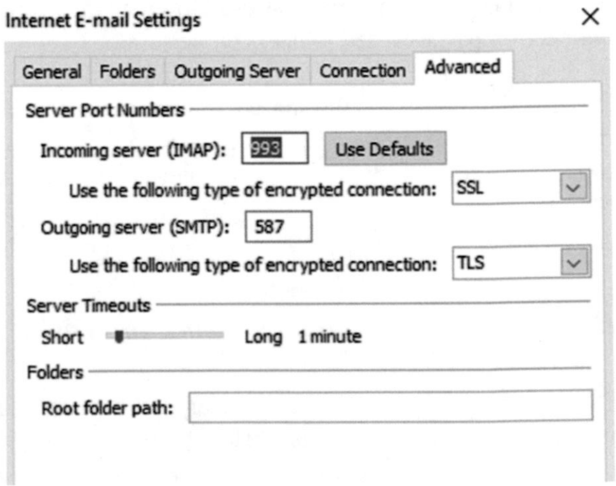

Fig. 5.12 Enable Microsoft BitLocker

Each block is encrypted differently based on the different tweak values. Additionally, in XTS each AES input is XORed with a different shifted version of the encrypted tweak. This can prevent an attacker from changing one specific bit in the disk.

It is argued then that while both CBC (undiffused) and XTS can prevent the first attacking type, XTS only is capable to prevent the second attacking type. This is one of the reasons for considering diffused CBC mode. XTS-AES is used by full-disk encryption systems, such as: LUKS, TrueCrypt, FileVault, and Microsoft BitLocker.

XTS incorporates two AES keys, K1 and K2. The first key: K1 is used for encrypting the sector number to compute per-block tweak values. The second key: K2 encrypts the actual data.

While XTS has its limitations and some performance issues, currently, it is considered as a very good choice.

FileVault for Apple OS/X

Apple Mac's FileVault uses XTS-AES-128 encryption that we described earlier. FileVault, used with Mac OS X Panther 10.3 and later, allows you upload a copy of your recovery key to Apple so you can recover your files via your Apple ID if you ever lose your password. This is optional if the user decides to store the key in iCloud.

FileVault 2 was introduced in Mac OS X 10.7 ("Lion"). Unlike FileVault 1, that encrypts only user data, FileVault 2 encrypts the whole disk Full-disk encryption (FDE) (Fig. 5.13).

One disadvantage of using FileVault 2 is that it uses the user's Mac OS password. It is generally recommended to use multi-factor authentication, rather than same OS credentials

Fig. 5.13 FileVault encryption in Apple

- TrueCrypt: This is a non-commercial freeware encryption application, for Windows, OS/X and Linux (www.truecrypt.org, truecrypt.sourceforge.net). The project started in 2004. TrueCrypt used LRW and XTS modes of encryptions. The project was terminated in 2014 and is no longer considered safe to use for disk encryption. The project faced some legal problems and disputes since it started. The disk encryption was abandoned from Windows operating systems beyond Windows XP, and similarly from recent Linux and Mac operating systems. Two alternative projects came to replace TrueCrypt: VeraCrypt and CipherShed.

VeraCrypt uses 30 times more iterations when encrypting containers and partitions in comparison with TrueCrypt. This means it takes more time to perform the encryption/decryption process.

VeraCrypt (https://veracrypt.codeplex.com/) has released version 1.21 in July 2017 with many security improvements in comparison with TrueCrypt. VeraCrypt uses XTS mode and supports AES encryption as well as some other algorithms: Serpent, Twofish, Camellia, and Kuznyechik.

In order to balance between security and performance, VeraCrypt supports parallelization in the encryption/decryption process.

- ESSIV: Encrypted salt-sector initialization vector (ESSIV) method for full-disk encryption is proposed to "replace" CBC mode. In particular, ESSIV proposed new method to generate Initialization Vectors (IVs). This is as the usual methods for generating IVs are predictable sequences of numbers based on time stamp or sector number which may permit certain attacks. IVs are generated from a combination of the sector number (SN) with the key hash. SN combination with the key in form of a hash makes the IV unpredictable. ESSIV is used by different versions of Linux since 2000.

Classical CBC may ensure confidentiality of the encrypted data but not integrity. If the plaintext is known to the attacker, it is possible to change parts of the plaintext block to specific values, while the blocks in between are changed to random values. Hence ESSIV can be effective in protections against attacks such as watermarking attacks in which specific parts of the data is tampered with or changed.

- Linux Unified Key Setup (LUKS): LUKS, created by Clemens Fruhwirth in 2004, is a popular Linux full disk encryption. The default encryption mode in early Linux versions is AES 256-bit, CBC mode. Recent versions switched to XTS due to the reasons we mentioned earlier. LUKS-encrypted disks can be used with LibreCrypt in Windows environment.
  Figure 5.14 shows command in Linux to display current encryption mode.

```
# cryptsetup luksDump /dev/sda5|grep Cipher
Cipher name:    aes
Cipher mode:    cbc-essiv:sha256
```

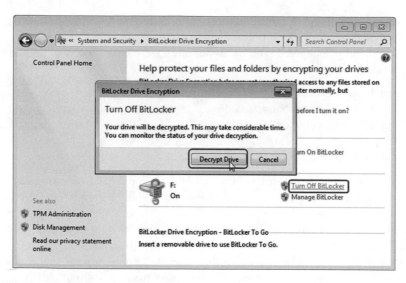

Fig. 5.14 An example of hash values generators

The command below shows how to use XTS mode:

```
cryptsetup luksFormat --cipher aes-xts-plain64 /dev/sdX
```

Information Integrity and Authentication Techniques and Methods

Information integrity techniques try to ensure that information is accurate, consistent, not changed, tampered with or exposed to unintended audience. While information integrity and hiding methods share similar goals, nevertheless, their processes and tasks can be used independently. In many applications that we mentioned earlier, hashing and information integrity is necessary to complete or complete the information encryption process.

Information integrity can have different implementations and applications in different domains or environments. Information trust and credibility are related terms to information integrity. For example, in the news, we can relate information/article with higher trust and integrity if reported by a credible news channel or website. In e-commerce, Certificate Authorities (CAs) are used to check the integrity of the content on a web-page and verify the identity of the owner.

This section can be largely divided, based in functions to integrity/manipulation checking and authenticity/originality checking.

- **Information integrity/manipulation**: For different telecommunication and information applications, it is important to verify that information is not changed, manipulated or tampered with by any unintended party. Those methods can be further divided general into data in stores and data in transit.

When data is stored in disks, databases, operating systems, etc. mechanisms exist to check that data is not changed. A typical method to do that is hashing. You can hash a certain text, a single file, a folder, a disk, etc. Figure 5.14 shows an example of hashing a simple text using different hashing algorithms.

We should notice the following characteristics for hashing algorithms:

- A hash value is a string of numbers and letters. The size of the string can vary from one hashing algorithm to another. For one hashing algorithm, the size of the hash output is fixed.
- As we can see from Fig. 5.14, different hashing algorithms will produce different hash values for the same original or input text, file, disk, etc.
- Hashing the exact same text using the same hashing algorithm should always generate the same hashing value. On the other hand, any slight change in the original text, (even addition/deletion of one character), should cause the hashing value to be different. The difference between original hash and new hash is not correlated with the level of addition or change on the original text. Those 3 criteria mention in this paragraph are keys to the hashing algorithm and process. In other words, this is the method to check information integrity. For example, forensic investigators use the hashing process to verify the integrity of the investigation process. The first thing they do when they acquire a possible evidence (e.g. a computer disk), is to hash the whole disk and record hash values (from different algorithms), record and attach it to the disk evidence. Next, they will get an image copy from that disk and start their investigation tasks on the image disk. In court, they will have to show original evidence and recorded hashing values along with current/image disk and recorded values to show that no single bit of data was added by them to the evidence.

Hash collisions may rarely occur. A hash collision can occur if two different files made the same hashing value. Hash functions should be "collision resistance" in general and this is one of the main required features in an accepted hashing algorithm. SHA-512 hashing algorithm generates large hash values in terms of size or number of characters in the hash output. This is one of the factors that make SHA-512 as one of the best hashing algorithms in terms of collision resistance".

- Unlike encryption, the hash value itself contains no information. You can't retrieve any information related to the original message from its associated hash value. Additionally, you can't reverse engineer the process and, somehow, retrieve the original message from the output hash value.

Here is another form or example of "collision resistance", let's assume that an adversary knows a message hash, so he/she will try hashing functions to find any text that will produce the same previous hash. Hashing functions or algorithms should not support this reverse process (i.e. to feed the hashing algorithm with a hash and expect to generate a message as an output). If we compare encryption with hashing, we can also say that encryption is a two-way function (encryption/decryption) while hashing is only one way (from message to message digest or hash value). It's infeasible to create the original content, knowing the hash value.

Information Authentication/Originality Checking

The "unique" hash value can either identify the message or the sender. For example, in digital certificates and e-commerce, and as users online communicate virtually, not physically, there is a need for them to be identified to each other virtually.

In PKI, asymmetric (e.g. RSA) algorithms are used to encrypt the message. This can protect the content of the message from being tampered or accessed by third parties between sender and receiver. Receiver also needs to be able to ensure that sender is the one who claim to be. The identity of the sender can be also spoofed. Digital certificates are created and signed by Certificate Authorities (CA) using hashing algorithms. Those signed digital certificates are sent with the encrypted messages (Fig. 5.15).

In Fig. 5.15 message digest refers to the hashing process output. The hash or the message digest is signed to produce digital signature. Hash function is used in this case for authentication, rather than message encryption. Receiver hashes the received message with the same hashing algorithm. If the new hash matches the received hash, message is considered identical.

Information Integrity/Authentication Applications

Message authentication
We described earlier the two main goals of hashing, message content originality or integrity verification and message sender authentication. There are many applications for both message content and sender verification. Those include ecommerce applications, remote and telecommunication services, email services and many others.

Fig. 5.15 PKI encryption with digital certificates

We already mentioned PKI and digital signatures/certificates as examples of applications using or utilizing hashing algorithms/tools. Next, we will describe examples of other applications:

Commitment schemes

There are many applications (e.g. bidding and gambling), where users need to commit to a value/number without revealing it until a certain transaction is completed. Party users can decide those values and hash them and make the hash values public. With the assumption that no collision is going to occur, there is a one-to-one relation between original number and hashed value which means that users will not be able to change the value afterward.

Password storage and verification

There is a need to have and keep many passwords for different applications by either individual users or enterprises. Storing those passwords in files as plain-texts is prohibited. Alternatively, hash values of those passwords can be stored. Even in cases of exposures for those hash values, attacker cannot retrieve original passwords, knowing their hash values. In one example of this, operating systems store users and their passwords (hash values) in certain folders or directories (Fig. 5.16).

ⓘ www.fileformat.info/tool/hash.htm

Results	
Original text	This is a test text
Original bytes	54:68:69:73:20:69:73:20:61:20:74:65:73:74:20:74:65:78:74 (length=19)
Adler32	424406db
CRC32	b172801e
Haval	cb5a98d31ee030fc0e0ad5997fecb23b
MD2	4b62f08a6eaa317511d16129fd4bb241
MD4	5c7398230367c2223b7250ce531706e7
MD5	734ab464c2b2c5803ff1b2d94729a6b5
RipeMD128	dfc49fe20a514af8474e0662d589d5c5
RipeMD160	1806fcd632d1f7c86236470974ed1cf55ea74e98
SHA-1	4d69b3afeeb3621d87935cd0f768360c63d1ec10
SHA-256	c62968ebcd6b7c706a8ac082db862682fa8be407106cb7eaa050c713a4e969d7
SHA-384	04d85d71fa44c6c60f2091c4301cd7084cbccb7dd1486cb99bd0cc0bdb0c102c1
SHA-512	eab1bca52815eff523f62243278a15f81a997b3843f9863f5d2cda1e7fd6f3f7343

Fig. 5.16 Ubuntu shadow content, a sample

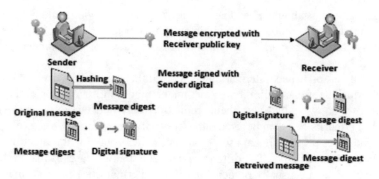

Fig. 5.17 Password hashing and verification

In addition to password storage, hashing can be used for password verification. To authenticate users, the password presented by the user is hashed and compared with the stored hash (Fig. 5.17). The process does not support the reverse process (i.e. get the password from the hash). Additionally, users cannot retrieve their passwords (but they can change it). Typically, such applications should allow users to try a finite number of tries for their password to prevent brute-force password cracking. Each password is often concatenated with a random, salt value before the hash function is applied.

Digital time-stamping

For digital forensics and many security functions, it is important to be able to find out the exact/accurate date/time of when a document was created, modified or accessed. Hash functions can be used to report events concisely, and to cause events based on documents without revealing their contents.

Digital certificates can be used to authenticate documents and verify the creator identity. Digital time stamping can be used to answer the question: "When this document was created or last modified?"

The originator of the document computes or creates the hash values and sends them in to the time-stamping service or Time Stamping Authority (TSA). The TSA concatenates a timestamp to the hash and calculates the hash of this concatenation.

The original document itself is only needed for verifying the timestamp. A time-stamping system cannot be compromised by the disclosure of a key because digital time-stamping systems do not rely on keys, or any other secret information.

Information Concealment/Authenticity Techniques, Methods and Applications

Unlike encryption, in steganography and watermarking, the goal is not to hide the original message or information from unintended audience. Their goal is related to copyright issues and hence, extra embedded text is added to the original data source (e.g. a file, an image, a video, etc.).

Steganography is about embedding a secret message/text within another message. This secret is carried innocuously within an innocent look. Traditionally, a watermark has been used to verify the authenticity/originality of a document, product, image, etc.

The main reason to use watermarking is "copyright issues". Producers of products such as: Music, software, film, book, etc. understand that many people may try to illegally use, copy or distribute their products (freely and without owner knowledge). Watermarks can then be used either to verify the actual owner or originator of the product or prevent illegal users from using the product without getting rid of the embedded watermarks. The difference between the two cases is that in the first case, watermarks are usually hidden (invisible watermarking) from normal users. Owners have their own methods to look for those watermarks for authenticity. Classical examples of this type include watermarks in some popular historical arts or paintings. Another popular watermarking example of this type is governments' watermarks on their money bills or currencies. Same thing applies to passports, bank notes or checks, etc.

In the second watermarking (visible watermarking) that we mentioned earlier, product originator intentionally wants this watermark to be visible. This is very popular in the software industry (also called software crippling, Fig. 5.18). Unlike physical products such as clothes, laptops, etc. software applications are easy to copy. Visible watermarking can help software providers distribute their software publicly (for marketing and demo). Watermarking does not prevent using or copying the software, but its software company proof of ownership. Users can see and evaluate software features. However, watermarks can prevent users from professionally utilizing the software (freely). Users are expected to upgrade to professional or commercial versions to get rid of the watermark. Visible watermarks serve similar purposes of "logos" and trademarks.

Steganography includes "hidden" messages within a larger text. Unlike watermarking, in steganography, hidden messages are not for copyright issues. In most watermarking cases, those messages are visible to users, while in steganography those are hidden messages that users will not notice. Users should not be able to detect steganography while they should not be able to delete or remove watermarking.

```
root@kali:~# cp /etc/shadow hash.lst
root@kali:~# more hash.lst
root:$6$Ye.heHsK$uhguCAA7ujTrSH/ldpqy/wLEeNIMMIR.4qYcAgB5aF1IY5h6VS5IOfa91IbibmX
LNwhwJmHGR05bXjk1Gmu.b.:16079:0:99999:7:::
daemon:*:16078:0:99999:7:::
bin:*:16078:0:99999:7:::
sys:*:16078:0:99999:7:::
sync:*:16078:0:99999:7:::
games:*:16078:0:99999:7:::
man:*:16078:0:99999:7:::
lp:*:16078:0:99999:7:::
mail:*:16078:0:99999:7:::
news:*:16078:0:99999:7:::
uucp:*:16078:0:99999:7:::
```

Fig. 5.18 (**a**) software watermark example. (**b**) A software watermark example (Copyright …)

For example, hackers use some tools to hide files (usually harmful) inside innocent files. Once victims receive and open those innocent files (e.g. using social engineering methods), embedded hidden files will be added or installed in the background. Usually those embedded messages or files will impact and increase original file size. However, some complex steganography methods can insert files with different sizes on other files without changing original file sizes (e.g. through file slack spaces, etc.). Watermarking can be visible or invisible while steganography is always invisible.

Steganography text can be added separately to original text. Alternatively, steganography can be added in the way original text is located out in the file. For example, in (Line-shift steganography), a text line is shifted certain size up to indicate (zero) or shifted the same size below to indicate 1.

The added text to the original information source should not cause any distortion. This is typically reported as part of the "robustness" of the watermarking or steganography process. Another related characteristic is called: "perceptibility" where users should not distinguish original from watermarked message. For example, users who are viewing an image that contains steganography should not see a distorted image or an image with quality problems. The image processing tools should be able to embed the hidden text or message while at the same time prevent human users from being able to notice it (by their naked eyes).

Steganography can be used in applications related to intelligent services and military applications. It is also used by several image processing applications, medical imaging, finger-printing applications, etc.

Skills Section

Use and Evaluate Encryption Algorithms

This exercise shows a simple Encryption performance test. Download and compile the Java code from: https://github.com/alsmadi/EncryptionsPerformanceTest

Once you successfully compile the code, run the code, choose the option (x) to run the code with different string size numbers. Records 5 different tests [with different integer values] and in each time record the time to encrypt for each one of the 3 encryption types. Summarize your results in a table. General instructions:

1. Download and Install JDK
   http://www.oracle.com/technetwork/java/javase/downloads/jdk8-downloads-2133151.html
2. Download and Install NetBeans
   https://netbeans.org/downloads/
3. Go to the link: https://github.com/alsmadi/EncryptionsPerformanceTest, click clone or download
4. In NetBeans, click [file - import project - from Zip]. You will get the code and build it.

Use and Evaluate Encryption Applications

- Using Windows operating system or image (beyond Windows Vista), follow the steps described in one of the following links to show the process of encrypting a disk in Windows. Make sure you document all your steps.

  (BitLocker Drive Encryption Step-by-Step Guide, link: https://technet.micro-soft.com/en-us/library/cc732725(v=ws.10).aspx)
  https://www.windowscentral.com/how-use-bitlocker-encryption-windows-10,
  https://docs.microsoft.com/en-us/windows/device-security/bitlocker/bitlocker-overview,
  https://www.howtogeek.com/192894/how-to-set-up-bitlocker-encryption-on-windows/).
  Alternatively, you can show the steps to do that either using Apple MAC or Linux operating systems (https://theintercept.com/2015/04/27/encrypting-laptop-like-mean/#osx, How to encrypt your Mac with FileVault 2, and why you absolutely should (http://www.macworld.com/article/2880039/how-to-encrypt-your-mac-with-filevault-2-and-why-you-absolutely-should.html).

- You will be asked to show how to encrypt and decrypt in Linux using LUKs. Follow the steps described in the link (http://www.forensicswiki.org/wiki/Linux_Unified_Key_Setup_(LUKS)) or similar ones (e.g. http://en.linuxreviews.org/Linux_Unified_Key_Setup, https://access.redhat.com/solutions/100463) to complete this task. Show also the commands to use for selecting different encryption modes of operations (e.g. CBC, XTS).

Use and Evaluate Information Integrity Algorithms

You are going to evaluate and compare different hashing algorithms (e.g. MD5, SHA, etc.). Collect first criteria to compare between those algorithms (e.g. performance, collision resistance, size, etc.). Then produce a table with different examples of input text to show results from each hashing algorithm.

Use and Evaluate Information Integrity Applications

There are some tools that claim that they can crack hashing algorithms (e.g. https://crackstation.net/, https://hashkiller.co.uk/md5-decrypter.aspx, http://www.md5online.org/). Use the website (http://www.fileformat.info/tool/hash.htm) to hash some texts. Then use hash crackers to evaluate the percentage of your text that can be successfully cracked.

Use and Evaluate Information Concealment Applications

Download image manipulation tool from (http://photodb.illusdolphin.net/en/download/). The tool can be used for steganography. Use the examples mentioned in the tool website to demo hiding and extracting files from images.

Experience/Ability Section

Select an information system/website, etc. and evaluate it from information reliability, and validity perspectives

OWASP website provides a cheat sheet related to the proper usage of TLS/SSL (https://www.owasp.org/index.php/Transport_Layer_Protection_Cheat_Sheet). Pick a website of your selection (that provides secure content). Then study all rules described in the sheet and check which rules are checked to exist in your tested website and which rules do not exist.

Conduct your own investigation about "cold boot attack" related to disk encryption. Make sure you cover what is "cold boot attack", how it occurs, on which encryptions, and methods suggested to prevent against it.

Select an information system/website, etc. and evaluate it from information privacy and protection perspectives

In this task, you will try to simulate an attack on a Linux disk. Follow the steps described in the link: Practical malleability attack against CBC-Encrypted LUKS partitions   (http://www.jakoblell.com/blog/2013/12/22/practical-malleability-attack-against-cbc-encrypted-luks-partitions/)

Some research references conducted experiments on different hashing algorithms and their resistance to collision. Many sorted the followings in ascending order: MD5, SHA1, SHA224, SHA256, SHA384, and SHA512. Conduct your own investigations/experiments to find out best algorithms in terms of "collision resistance". We showed in the text one factor that impacts this issue (The size of the hash output). Show if there are other factors that can also affect this issue.

There are different types of attacks on hashing algorithms. Examples of those attacks include: Guessing attack, Birthday attack, precomputation of hash values, Long-message attack for 2nd preimage, etc. Select one type of those attacks and research on how to implement and evaluate such attack.

USA NIST conducted a hash competition to evaluate hashing algorithms over the years from 2004–2007 (http://csrc.nist.gov/groups/ST/hash/sha-3/pre-sha-3-comp.html). Summarize goals, acceptable requirements, experiments and main proposed algorithms, findings as the results of this competition.

Questions

- Compare between 3 different encryption algorithms in terms of performance.
- What is the current best encryption algorithm and why?

- What is the current best hashing algorithm and why?
- Make a comparison between symmetric and asymmetric encryption.
- Which is the current best block cipher to use and why?
- How can you best test an encryption algorithm?
- What is the difference between TSL and SSL? Which one is currently more in use?
- Compare between 3 different hashing algorithms in terms of performance.
- What are the main characteristics to decide if an encryption algorithm is effective?
- What is the difference between invisible watermarking and steganography?

References

Goldwasser, S., & Bellare, M. (2008). *Lecture notes on cryptography*. Summer course on cryptography, MIT, 1996–2001.

http://aesencryption.net/

http://extranet.cryptomathic.com/rsacalc/index

http://searchsecurity.techtarget.com/definition/RSA

http://www.cs.mtu.edu/~shene/NSF-4/RSA-Downloads/index.html

https://www.cs.drexel.edu/~introcs/Fa11/notes/10.1_Cryptography/RSA_Express_EncryptDecrypt.html

https://www.cs.drexel.edu/~jpopyack/IntroCS/HW/RSAWorksheet.html

https://www.di-mgt.com.au/rsa_alg.html

https://www.tools4noobs.com/online_tools/encrypt/

https://www.tutorialspoint.com/cryptography/

https://www.xarg.org/tools/caesar-cipher/

Chapter 6
Network Security

Overview

Providing appropriate protection techniques is significant to combat cyber-threats and preserve the integrity, confidentiality and availability of computer networks. The increasing volume of malicious cyber-attacks demands a collaborative effort between security professionals and researchers to design and develop effective cyber-defense systems. Cyber-attacks continue to rise worldwide in a manner that costs companies millions of dollars each year and leads to loss or misuse of information assets. Therefore, companies are under pressure to avoid the occurrence of these attacks or decrease the damage they cause to computer networks. In this chapter we will cover the aspects of network security, the types of network attacks and the techniques for measuring network security.

Knowledge Sections

In the knowledge section, we will cover the OSI (Open Systems Interconnection) layers and the types of attacks that target each one of them.

Open Systems Interconnection (OSI)

The Open Systems Interconnection model (OSI Model) describes the standards need for communications in computer networks without focusing on the internal structure. This model consists of many layers as shown in Fig. 6.1. In the following section we will focus on the types of protocols used in each layer.

© Springer International Publishing AG 2018
I. Alsmadi et al., *Practical Information Security*,
https://doi.org/10.1007/978-3-319-72119-4_6

OSI MODEL	
7	**Application Layer** Type of communication: E-mail, file transfer, Client/server.
6	**Presentation Layer** Encryption data conversion: ASCN to EBCDIC, BCD to binary, etc.
5	**Session Layer** Starts, stop session, Maintains order.
4	**Transport Layer** Ensures delivery of entire file or message.
3	**Network layer** Routes data to different, LANs and WANs based, On network address.
2	**Data Link (MAC) Layer** Transmits packets from node to node based on station address.
1	**Physical layer** Electrical signals and cabling

Fig. 6.1 The layers in the OSI Model

In the following section we will focus on the types of protocols used in each layer, which are also summarized in Fig. 6.2.

Application Layer

TFTP (Trivial File Transfer Protocol) — Send file quickly
DNS (Domain Naming System) — Translate between the domain name and IP
DHCP (Dynamic Host Configuration Protocol) — Assign IP addresses.
Telnet— Remote connection.
HTTP (Hypertext Transfer Protocol) —Browsing web pages.
FTP (File Transfer Protocol) — reliable file transfer.
SMTP (Simple Mail Transfer Protocol) — email transfer.
POP3 (Post Office Protocol v.3) —Email retrieval.

Transport Layer

Transmission Control Protocol
TCP is a connection oriented and reliable protocol. TCP is reliable since it gives sequence numbers to each packet that sent during the connection session. In addition, a delivery notice (acknowledgement) is used to confirm that the receiver has

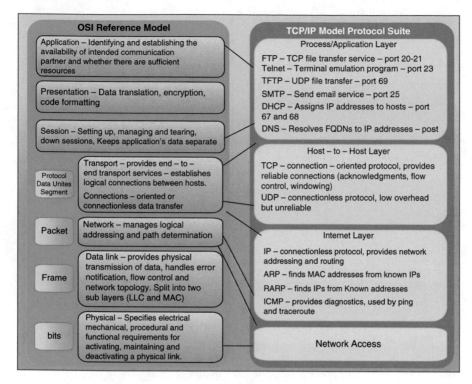

Fig. 6.2 Some protocols in the OSI Model

received. While the delivery of packets is guaranteed using this method, it costs more to use in data communications.

User Datagram Protocol

As opposed to TCP UDP is a connection less protocol. There is no guarantee for the packet delivery process. While there is a good chance that the data will be delivered, there is probability that the data will be lost in its way, this is since the communication using this protocol is considered cheap compared to TCP. The Communication using UDP does not mean that the data will not reach to its destination, however there is a chance that some packets might not be delivered.

Connection establishment process of TCP is done using Three-way handshake and the following steps (summarized in Fig. 6.3)

SYNx (Synchronize sequence number)
SYNy, ACKx + 1 (Acknowledgement)
ACKy + 1, data

The process requires allocation of memory resources as shown in Fig. 6.3 during TCP connection establishment. Resources has to be allocated by both endpoints for information related with connection

Fig. 6.3 Connection
establishment process of
TCP

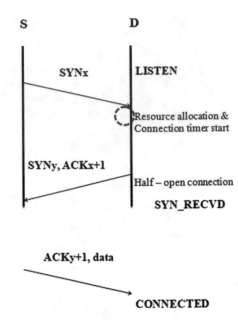

Network Layer

Providing IP addresses is the main responsibility of this layer. Devices such as routers work at this layer. There are several functionalities for this layer such as finding
the best route for packets to reach their destination, and connecting Ethernet, Token
Ring and other datalink types together.

Types of the Addresses in the Internet

There are several types of addresses in the communication networks including

Media Access Control (MAC)
Associated with the network interface card (NIC)
It can be 48 bits or 64 bits
IP addresses
Associated with TCP/IP
32 bits for IPv4 and 128 bits for IPv6
E.g., 127.3.21.5
IP addresses + ports for the transport layer
E.g., 122.1.32.8:60
Domain names for the application/human layer
E.g., www.purdue.edu

Figure 6.4 shows an example on each form of addresses

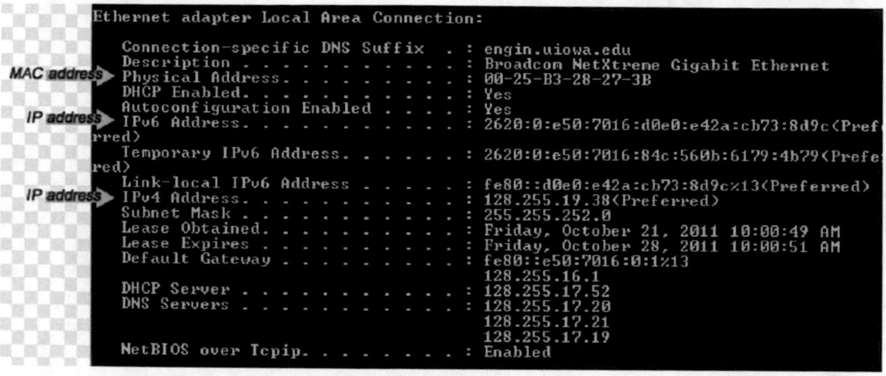

```
         Ethernet adapter Local Area Connection:

               Connection-specific DNS Suffix  . : engin.uiowa.edu
               Description . . . . . . . . . . . : Broadcom NetXtreme Gigabit Ethernet
MAC address►   Physical Address. . . . . . . . . : 00-25-B3-28-27-3B
               DHCP Enabled. . . . . . . . . . . : Yes
               Autoconfiguration Enabled . . . . : Yes
IP address►    IPv6 Address. . . . . . . . . . . : 2620:0:e50:7016:d0e0:e42a:cb73:8d9c(Pref
               rred)
               Temporary IPv6 Address. . . . . . : 2620:0:e50:7016:84c:560b:6179:4b79(Prefe
               red)
               Link-local IPv6 Address . . . . . : fe80::d0e0:e42a:cb73:8d9c%13(Preferred)
IP address►    IPv4 Address. . . . . . . . . . . : 128.255.19.38(Preferred)
               Subnet Mask . . . . . . . . . . . : 255.255.252.0
               Lease Obtained. . . . . . . . . . : Friday, October 21, 2011 10:00:49 AM
               Lease Expires . . . . . . . . . . : Friday, October 28, 2011 10:00:51 AM
               Default Gateway . . . . . . . . . : fe80::e50:7016:0:1%13
                                                   128.255.16.1
               DHCP Server . . . . . . . . . . . : 128.255.17.52
               DNS Servers . . . . . . . . . . . : 128.255.17.20
                                                   128.255.17.21
                                                   128.255.17.19
               NetBIOS over Tcpip. . . . . . . . : Enabled
```

Fig. 6.4 Types of Addresses in Internet

Routing and Translation of Addresses

The translation between addresses during the connection sessions include

Translating between IP and MAC addresses which is done using the Address resolution
protocols for IPV4 and Neighbor Discovery Protocol (NDP) for IPv6
Routing using the IP addresses for network packets and connections
Translating domain names and IP addresses

Network Attacks at Different Layers

Eavesdropping
In this type of attack the attacker gains access to the data path in the network to
listen for the conversions or investigate the traffic. The major reason of this attack is
that the majority of communications in the network occur in an unsecured and clear
text. This type of attack is also known as sniffing or snooping. An example of this
type of attacks is given in Fig. 6.5. The Eaves dropping attack is considered one of
the major security problem that administrators usually tackle in the network.
   How to solve the eavesdropping attack:

* **Encryption:** It is considered one of the major defense mechanisms against
  eavesdropping. Systems and application data can be encrypted using the existing
  encryption mechanisms
* **Network segmentation:** In this type of countermeasures the network is divided
  into many segments. Then some access control techniques are used to authorize
  specific users in each segment.

Network access control (NAC): NAC ensures that network devices are trusted
before the network connectivity is allowed.

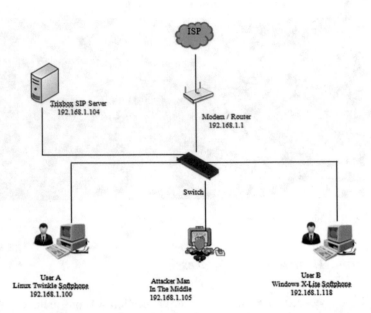

Fig. 6.5 An example on the Eavesdropping attack (http://blog.nostratech.com/2015/09/keamanan-jaringan-serangan-proteksi_18.html)

Physical security: Prevent access to network points that offer direct connectivity to the corporate network.

Data Modification Attacks

In this type of attack, the attacker attempts to change the data transmitted over communication networks. The data stream is modified to perform deception of one entity as some other. The attacker may also try to modify messages in transit initiate denial of service attacks and other types of attacks such as Man in the Middle (MiM) attack, etc.

Example on data modification attacks is the SQL injection as shown in Fig. 6.6 where the attacker may write suspicious queries to update particular records in the database.

Identity Spoofing Attacks

This protocol is used in the translation between the IP addresses and MAC addresses. This is needed to be done for NIC to send a packet. Each host maintains a table of IP to MAC addresses. The ARP process itself consists of two types of messages ARP request and ARP reply.

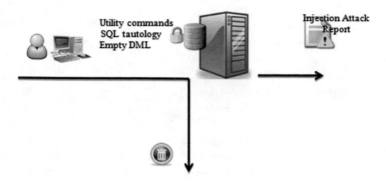

Fig. 6.6 An example on data modification attack (https://www.enterprisedb.com/sql-injection-attacks-rise)

Fig. 6.7 ARP protocol

Computer A
192.168.1.1
Mac: AAA

Computer B
192.168.1.2
Mac: BBB

ARP Spoofing Attacks

Network addressing and routing leads to several concrete security problems including

ARP is not authenticated and this could lead to APR spoofing (or ARP poisoning)
Network packets pass by un-trusted hosts which may lead to Packet sniffing
DNS is not authenticated which may lead to DNS poisoning attacks

The computer that sends packets needs to know the MAC address for the destination computer. Therefore, the ARP protocol is needed. Let's show an example as shown in Fig. 6.7. The connection is initiated in the following steps

1. A packet with its source and destination IP addresses is created
2. The packet header will also include the source and destination MAC addresses in an Ethernet frame
3. The source computer will need to send an ARP request to get the MAC address of the destination computer. This is where the ARP request is sent

Ping utilizes the ICMP protocol. The IP that is associated with packet is 192.168.1.1 the destination IP on the other hand is 192.168.1.2 (Fig. 6.8). The next step is to encapsulate the source and destination MAC addresses with the packet in an Ethernet frame.

```
C:\\Users\\ComputerA>ping 192.168.1.2
Pinging 192.168.1.2 with 32 bytes of data:
Reply from 192.168.1.2: bytes=32 time=15ms TTL=57
Reply from 192.168.1.2: bytes=32 time=15ms TTL=57
Reply from 192.168.1.2: bytes=32 time=14ms TTL=57
Reply from 192.168.1.2: bytes=32 time=17ms TTL=57

Ping statistics for 192.168.1.2:
Packets: Sent = 4, Received = 4, Lost = 0 (0% loss),
Approximate round trip times in milli-seconds:
Minimum = 14ms, Maximum = 17ms, Average = 15ms
```

Fig. 6.8   Ping Command

```
C:\\Users\\ComputerA>arp -a

Interface: 192.168.1.1 --- 0xb
   Internet Address        Physical Address      Type
   192.168.1.2             00-0c-29-63-af-d0      dynamic
   192.168.1  .255         ff-ff-ff-ff-ff-ff      static
   224.0.0.22              01-00-5e-00-00-16      static
   224.0.0.252             01-00-5e-00-00-fc      static
   239.255.255.250         01-00-5e-7f-ff-fa      static
   255.255.255.255         ff-ff-ff-ff-ff-ff      static
```

Fig. 6.9   ARP Command

The computer A knows the IP address for computer B, however the MAC address of computer B is not known. The Address Resolution Protocol (Fig. 6.9) is used to resolve this issue.

The ARP table is initially empty, therefore the MAC address for computer B is not known. As such computer A will initiate an ARP request (Fig. 6.10). The ARP request will initiate a query about the MAC address of: 192.168.1.2. The broadcast MAC address (FF:FF:FF:FF:FF:FF) is used. This query will be forwarded to all computers in the Network.

When the broadcast message received, an ARP reply message from the destination computer is initiated with the MAC address for that computer. The ARP table for computer A will be updated to include the MAC address for computer B.

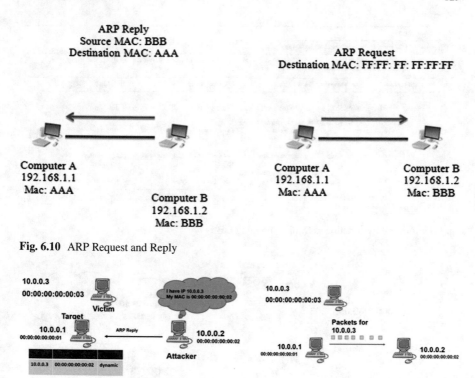

Fig. 6.10 ARP Request and Reply

Fig. 6.11 ARP Spoofing Attack

Address Resolution Protocol (ARP)

The mechanism is called Address resolution protocol. IP to MAC mappings are stored in each device's ARP table (the phone book in your analogy). To simplify: In most cases, to resolve the MAC address associated with an IP address, you send a broadcast ARP packet (to all devices in the network), asking who has that IP address. The device with that IP address replies to the ARP (with its MAC address).

The ARP spoofing attack is initiated by spoofing an ARP message to an Ethernet LAN to have other machines associate IP addresses with the attacker's MAC (Fig. 6.11).

Countermeasures

Using Router's static ARP tables
   [edit interfaces]fxp0 {unit 0 {family inet {address 10.10.0.11/24 {ARP 10.10.0.99 MAC 0001.0002.0003;arp 10.10.0.101 MAC 00:11:22:33:44:55 publish;}}}}

	Command	Purpose
Step 1	Switch# show ip dhcp snooping binding	Views the DHCP snooping database
Step 2	Switch# ip dhcp snooping binding binding-id vlan vlan-id interface interface expiry lease-time	Adds the binding using the 'ip dhcp snooping' exec command
Step 3	Switch# show ip dhcp snooping binding	Checks the DHCP snooping database

This example shows how to manually add a binding to the DHCP snooping database:

```
Switch# show ip dhcp snooping binding
MacAddress          IpAddress        Lease(sec)   Type            VLAN  Interface
------------------  --------------   ----------   -------------   ----  --------------------
Switch#
Switch# ip dhcp snooping binding 1.1.1 vlan 1 1.1.1.1 interface gi1/1 expiry 1000

Switch# show ip dhcp snooping binding
MacAddress          IpAddress        Lease(sec)   Type            VLAN  Interface
------------------  --------------   ----------   -------------   ----  --------------------
00:01:00:01:00:01   1.1.1.1          992          dhcp-snooping   1     GigabitEthernet1/1
Switch#
```

Fig. 6.12 Adding Information to DHCP Snooping DB

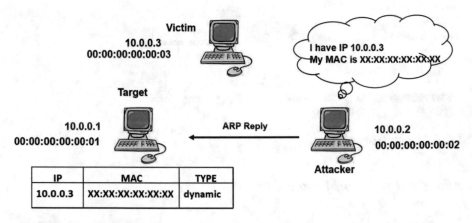

Fig. 6.13 ARP-based attacks: Denial of service Attack

1. DHCP Snooping utilizes a filtering mechanism to filter-out the untrusted messages using a binding table (Fig. 6.12). Those messages are received from outside firewall and can cause denial of service attacks. The binding table contains the MAC, IP addresses and Interface information.

Denial of Service Attack

A malicious entry with a non-existent MAC address can lead to a DOS attack (Figs. 6.13 and 6.14)

Victim will be unable to reach the IP for which the forged packet was sent by the attacker

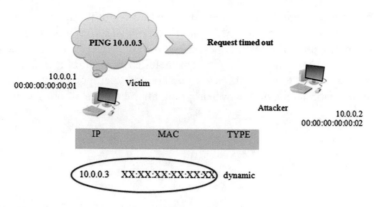

Fig. 6.14 ARP-based attacks: Denial of Service

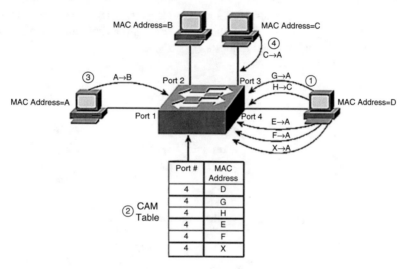

Fig. 6.15 Example on MAC Flooding Attacks

MAC Flooding Attacks

MAC flooding attack is a network-based attack in which the attacker initiates too many Ethernet frames from invalid MAC address. The CAM table overflow or MAC flooding attack aims to restrict the possibility of adding more MAC addresses to the CAM table. Traffic monitoring is yet another reason for this type of attacks.

An example on MAC flooding attack is given in the Fig. 6.15. Port 4 is where the compromised machine is attached. A fake MAC addresses G, H, E and F are used to send packet on port 4. The Compromised machine's MAC address is D.

The frames that are sent on port 4 cause the CAM table to fill up. The CAM table will not be able to learn any new MAC addresses. In addition, it will be unable to perform port mapping. Port 1 is associated with MAC address A. Since the CAM table is full, the MAC address for A is not included; therefore, if another host C wants to send a frame to A, the frame will be flooded to all ports in the LAN. This may lead to A DoS attack or an eavesdropping attempt since all ports will receive the frame sent.

A TCP SYN Flooding

TCP SYN flooding is a denial of service attacks in which SYB request are initiated to consume server resources which leads to unresponsive commands. An example on this type of attacks is given in Fig. 6.16. Initially the attacker sends many connection requests using a spoofed source addresses. The victim machine will allocate resources per request. The requests are kept until a time out which leads to exhausted resources and rejected requests.

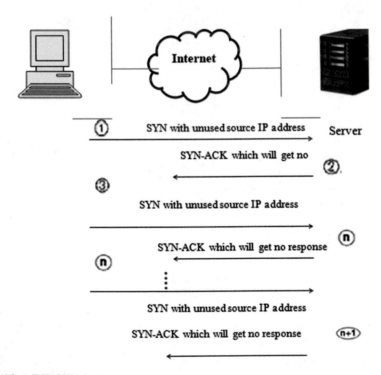

Fig. 6.16 A TCP SYN flood

Denial of Service Attacks (Smurf Attack)

Using a subnet broadcast address to send too many ICMP requests leads to SMURF attacks. The attacks occur if the following steps are done correctly (Fig. 6.17)

1. A spoofed address sends an echo request to an IP broadcast address
2. An echo reply is sent to the victim

A SMURF attack consists of sending ping request to broadcast address (X.X.X.255) of a router that transmits ICMP request to other devices behind it. The attacker computer used a spoofed source address that belongs to the victim. Since the handshake process is not included in the ICMP command the destination has no way of proving the identity of the Source IP address. The router will transmit the request to other devices. Those devices will respond back to the ping command to the victim host instead of the attacker host, which leads to SMURF attack.

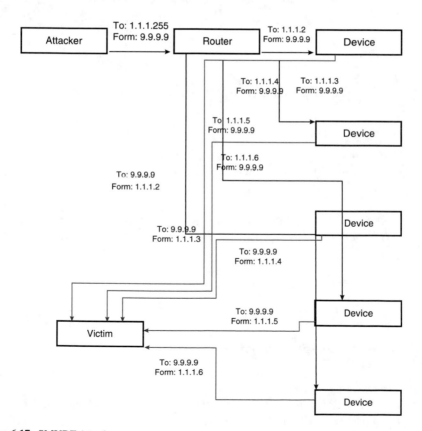

Fig. 6.17 SMURF Attack

Packet Sniffing Attacks

Packet sniffers are used to capture network packets then investigate the content that is sent as a plain text. There are several protocols that send information in plain text including

- FTP
- Telnet
- SNMP
- POP

Sniffers can be used to initiate several attacks such as traffic analysis to gain information and then corrupt the confidentiality of information

There are several mechanisms that can be used to mitigate sniffing attacks such as

- **Anti-Sniffing tools:** Those tools can be used to discover any sniffers
- **Cryptography:** This technique is not used as a sniffer detection mechanism, instead it is used to minimize the negative effects of the sniffers

Skills Section

Network Traffic Analysis Using Wireshark and Other Tools

- Tcpdump

    - Unix-based command-line tool used to intercept packets
    - Including filtering to just the packets of interest

- Tshark

    - Tcpdump-like capture program that comes w/ Wireshark
    - Very similar behavior & flags to tcpdump

- Wireshark

    - GUI for displaying tcpdump/tshark packet traces

Tcpdump installation
yum install tcpdump

Capture traffic from specific interface
tcpdump -i eth0

Tcpdump example
01:46:28.808262 IP danjo.CS.Berkeley.EDU.ssh > adsl-69-228-230-
7.dsl.pltn13.pacbell.net.2481: 2513546054:2513547434(1380) ack 1268355216
01:46:28.808271 IP danjo.CS.Berkeley.EDU.ssh > adsl-69-228-230-

7.dsl.pltn13.pacbell.net.2481: P 1380:2128(748) ack 1 win 12816

01:46:28.808276 IP danjo.CS.Berkeley.EDU.ssh > adsl-69-228-230-

7.dsl.pltn13.pacbell.net.2481: 2128:3508(1380) ack 1 win 12816

01:46:28.890021 IP adsl-69-228-230-7.dsl.pltn13.pacbell.net.2481 >

danjo.CS.Berkeley.EDU.ssh: P 1:49(48) ack 1380 win 16560

Tshark Example

1190003744.940437 61.184.241.230 - > 128.32.48.169 SSH Encrypted request
packet len = 48

1190003744.940916 128.32.48.169 - > 61.184.241.230 SSH Encrypted response
packet len = 48

1190003744.955764 61.184.241.230 - > 128.32.48.169 TCP 6943 > ssh [ACK]
Seq = 48

Ack = 48 Win = 65,514 Len = 0 TSV = 445871583 TSER = 632535493

1190003745.035678 61.184.241.230 - > 128.32.48.169 SSH Encrypted request
packet len = 48

Wireshark Example

- Packet sniffer/protocol analyzer (Fig. 6.18)
- Open Source Network Tool

Capture Options: Promiscuous mode is used to capture all traffic (Fig. 6.19).

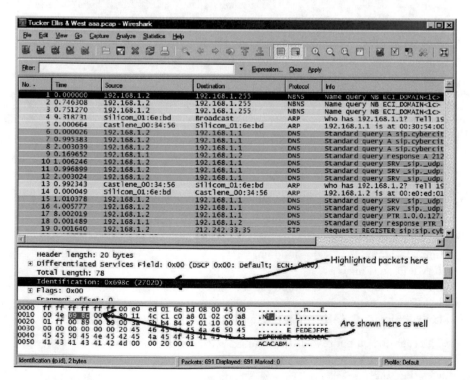

Fig. 6.18 Wireshark traffic capturing software http://www.netresec.com/?page=
Blog&month=2016-11&post=BlackNurse-Denial-of-Service-Attack

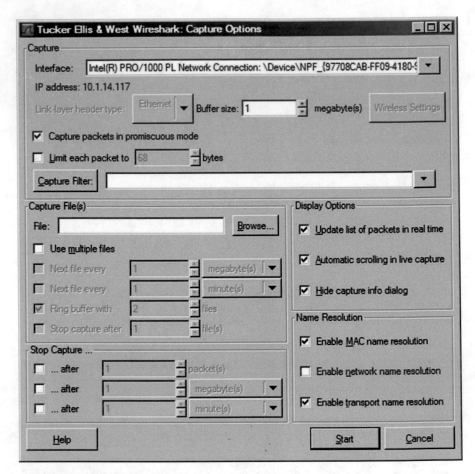

Fig. 6.19 Evaluation of Denial-of-Service Attacks Using Wireshark http://www.netresec. com/?page=Blog&month=2016-11&post=BlackNurse-Denial-of-Service-Attack

In Linux Command line we can evaluate how effective ICMP type 3 attacks. In this setup you can use hping3 to send ICMP floods like this:

- ICMP net unreachable (ICMP type 3, code 0):
  hping3 --icmp -C 3 -K 0 --flood [target]
- ICMP Echo (Ping):
  hping3 --icmp -C 8 -K 0 --flood [target]
- ICMP Echo with code 3:
  hping3 --icmp -C 8 -K 3 --flood [target]

The capinfos output reveals that hping3 was able to push a 303.000 packets per second (174 Mbit/s), which is a way to overload a network device vulnerable to DoS attacks (Fig. 6.20).

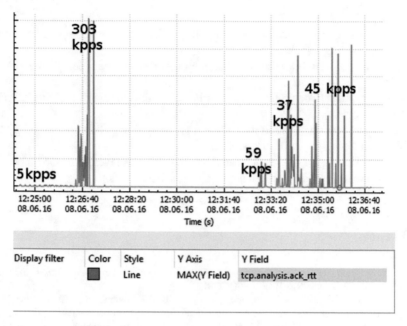

Fig. 6.20 Wireshark IO graph http://www.netresec.com/?page=Blog&month=2016-11&post= BlackNurse-Denial-of-Service-Attack

Applications Section

- **Applications for Encryption Techniques for Network security.**

  You are expected to select specific encryption algorithm, implement it in C++ or Java and describe why data encryption is important for Network Security

- **Applications of Network Protocols According to security Threats**

  You are expected to classify all network protocols in each OSI layer based on the attacks that may target each protocol. Summarize your findings:

  – List of protocols per layer.
  – Types of vulnerabilities in each protocol.
  – Types of attacks that target each protocol

- **Applications for detecting Network Attacks using Traffic Analysis**

  You are expected to select one of the packet capturing applications (e.g. Wireshark). Visit many websites.

- Check if the site use any encryption in submitting your data(e.g., username and password)
- Check what fields are encrypted when submitting your data

Questions
- In the scope network security, describe the differences between ARP Spoofing and SynFlooding attacks.
- In the scope of network security, describe the difference between DoS and Man in The middle attacks.
- In the core of network Security, Describe the main vulnerabilities in UDP and TCP protocols.
- Make a comparison between the different types of attacks that are caused by the vulnerabilities in the ARP protocol.
- Describe examples of threats that can target the Network layer in the OSI model.
- What is the difference between active and passive attacks?

References

Barford, P., Duffield, N., Ron, A., Sommers, J. (2009). *Network performance anomaly detection and localization*. Proc. 2009 IEEE INFOCOM.
Bertsekas, D., & Gallagher, R. (1991). *Data networks* (2nd ed.). Englewood Cliffs: Prentice Hall.
Bishop, M. (2003). *Computer security: Art and science*. Boston: Addison Wesley.
CERT. Advisory 2001–09: Statistical Weaknesses in TCP/IP Initial Sequence Numbers. http://www.cert.org/advisories/CA-2001-09.html
CERT. Advisory CA-96.21: TCP SYN Flooding and IP Spoofing Attacks. http://www.cert.org/advisories/CA-1998-01.html
Chiu, D., & Jain, R. (1989). Analysis of the increase and decrease algorithms for congestion avoidance in computer networks. *Computer Networks and ISDN Systems, 17*(1), 1–14. http://www.cs.wustl.edu/~jain/papers/cong_av.htm.
Cisco Systems Inc.. How NAT Works. http://www.cisco.com/en/US/tech/tk648/tk361/technologies_tech_note09186a0080094831.shtml
White paper.. *Understanding IP addressing: Everything you ever wanted to know*. http://www.3com.com/other/pdfs/infra/corpinfo/en_US/501302.pdf

Chapter 7
Web and Database Security

Overview

Security in web applications is the most important concern when it comes to processing transactions in the web. One of the major issues is the security and privacy of data and information transferred, stored and processed through at real time. These days, many online transactions between client and server are executed at the cloud data centers, where such sensitive data run on virtual resources. Like Several other systems, web-based systems, Cloud Web applications are vulnerable and proned to various types of web Injection attacks which result from transferring untrusted content from web to the server side so a secure communication should be satisfied to prevent web security threats. This chapter will introduce the types of attacks that target web applications. In addition, several examples on many attack scenarios are introduced.

Knowledge Section

In the knowledge section, we will cover the types of web and database attacks and many other scenarios on those attacks.

Web Applications

Those days the big trend is to offer software services on the web. Many services such as online banking, shopping, government, bill payment, tax prep, customer relationship management. In general, application code split between client and server.

© Springer International Publishing AG 2018   139
I. Alsmadi et al., *Practical Information Security*,
https://doi.org/10.1007/978-3-319-72119-4_7

Client code executes Java script. Server code includes, PHP, Ruby, Java, or ASP. Security is rarely the main concern when it comes to web applications and vulnerabilities such as poorly written scripts, inadequate input validation, and inadequate protection of sensitive data are the main reasons for web attacks.

The Two Sides of Web Security

1. Web browser: The software that is used to browse website and it can be attacked by any web site it visits. Attacks on Web browser may results in Malware installation (keyloggers, Botnets, Document theft from corporate network and Loss of private data)
2. Web application code: this code runs at web site, e.g. Banks, e-merchants, blogs and it can be written in PHP, ASP, JSP, and Ruby. There are many potential bugs that leads to stolen and defaced sites, such as

   (a) Cross Site Scripting(XSS): Malicious code injected into a trusted context (e.g., malicious data presented by a trusted website interpreted as code by the user's browser)
   (b) Cross Site Request Forgery(XSRF): bad website forces the user's browser to send a request to a good website
   (c) SQL injection: Malicious data sent to a website is interpreted as code in a query to the website's back-end database

In the following section, we will focus on the types of those attacks

Web Threat Models

There are several threat models to initiate attacks. This can happen at three levels, web-attacks, network attacks and malware attacks.

- Web attacker: Usually the objective here is to control the target website, obtain SSL/TLS certificate for the targeted website, or force users to visit the targeted website.
- Network attacker: in this form, the attacks can be categorized as active or passive. Example of passive attacks is the wireless Eavesdropper. Example on active attacks is the DNS poisoning
- Malware attacker: In this type of attacks the Attacker escapes browser sandbox

Cross Site Scripting

XSS can be initialized using different techniques, for instance, the attacker may inject malicious code into a benign website by loading it onto a valid server as part of a client review or a web-based email. Alternatively, the code may be injected into a URL and sent to user as an email When the user taps the URL, the content will be transmitted to the benign sever and then returned as part of a request of user credentials.

In Table 7.1a, a phishing email asks the user to click on a URL. The URL is not human-readable, but it can be translated into a readable form after mapping the hexadecimal characters, as shown in Table 7.1b. Then, the JavaScript code embedded into the search query will be executed upon visiting the target website, which will inject the HTML code (fetched from www.very-bad-site.com) into the code the user's browser would normally render.

In the cross site scripting attack, malicious markup and script is entered in the web pages that are viewed by other users. If proper care is not taken to filter this malicious piece of markup, the script gets stored in the system and also rendered on web pages.

Another example on XSS is given in the Figs. 7.1 and 7.2. As noticed in Fig. 7.1, the user can enter any text in the textbox and the textarea, including HTML markup

Table 7.1 (a) The URL sent in a Phishing email. (b) Translating URL into human readable form

http://www.well-known-financial-institution. com/?q=%3Cscript%3Edocument.write%28%22 %3Ciframe+src%3D%27http%3A%2F%2F www.very-bad-site.com%27+FRAMEBORDER %3D%270%27+WIDTH%3D%27800%27+HEI GHT%3D%27640%27+scrolling%3D%27auto% 27%3E%3C%2Fiframe%3E%22%29%3C%2Fsc ript%3E&...=...&...	http://www.well-known-financial-institution. com/?q=<script>document.write("<iframe src='http://www.very-bad-site.com' FRAMEBORDER='0'WIDTH='800' HEIGHT='640' scrolling='auto'></ iframe>")</script>&...=...&...">

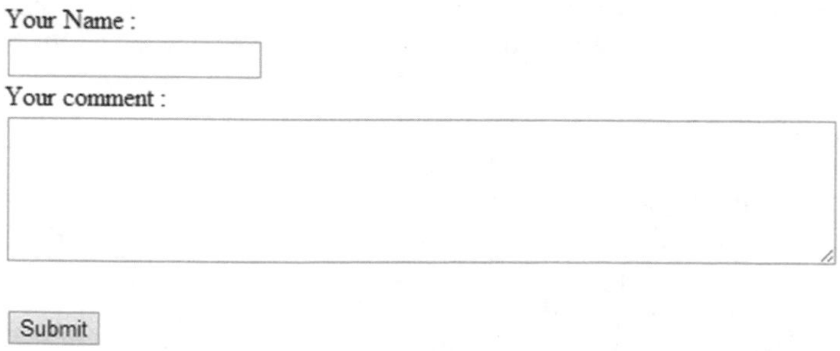

Your Name :

Your comment :

Submit

Fig. 7.1 Textbox and text area in a web form

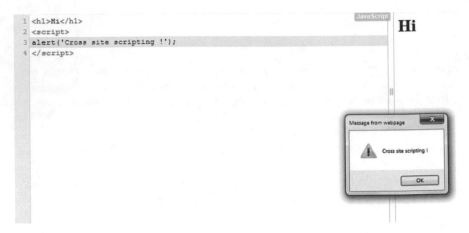

Fig. 7.2   Textbox and text area in a web form

Fig. 7.3 Cookies
associated with a specific
webpage

```
HTTP/1.1 303 See Other
Location: www.facebook.com
Set-Cookie: user=me@yegor256.com; Secure; HttpOnly
```

Fig. 7.4 Sever response
when the cookie header is
correct

```
HTTP/1.1 303 See Other
Location: www.facebook.com
Set-Cookie: user=me@yegor256.com
```

tags and script fragments! Once the form is submitted the posted data is saved in the database. The form is submitted to the SaveData action method. The SaveData () saves the data in a SQL Server database table named Comments.

Now assume that a use enters the following text in the comments textarea in Fig. 7.2:

When such a user posts the above content it gets saved in the database. Later when this saved content is rendered on a web page it executes the script. There are yet other types of XSS attack that is associated with cookie-based authentication, based on a browser's ability to expose all cookies associated with a web page to JavaScript executed inside it (see Fig. 7.3).

An attacker may inject some malicious JavaScript code into the page; this will happen only if your entire HTML rendering is done wrong, and this code will gain access to the cookie. Then, the code will send the cookie somewhere else so the attacker can collect it. To prevent this, there is another flag to prevent this type of attacks. The presence of this flag will tell the browser that this particular cookie can be transferred back to the server only through HTTP requests. JavaScript won't have access to it. The server matches the provided information with its records and decides what to do (Fig. 7.4). If the information is invalid, it returns the same login

```
HTTP/1.1 303 See Other
Location: www.facebook.com
Set-Cookie: user=b1ccafd92c568515100f5c4d104671003cfa39
```

Fig. 7.5 Cookie encryption to prevent XSS

Fig. 7.6 A web page on which is vulnerable to CSRF

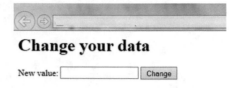

page, asking you to enter it all again. If the information is valid, the server returns something like this:

If the server trusts any browser request with a user email in the Cookie header, anyone would be able to send my email from another place and get access to my account. The first step to prevent this is to encrypt the email with a secret encryption key, known only to the server (Fig. 7.5). Nobody except the server itself will be able to encrypt it the same way the server needs to decrypt it. The response would look like this, using an example of encryption by XOR cipher with bamboo as a secret key:

XSS has several impacts

1. Several types of attacks including Denial-of-service attacks, crash users browser, pop-up-flooding, redirection
2. Access to Users' machine using ActiveX objects to control machine and uploading local data to attacker's machine
3. Spoil public image of company, Load main frame content from other locations, and redirect to dialer download

Cross Site Request Forgery

CSRF is yet another type of injection attacks that can be initiated as part of phishing campaigns. The attacker sends emails to victims to lure them into visiting a web page that is under attacker control (Blatz 2007, Nagar and Suman 2016). The attacker hides several executable elements in his page (e.g., Java scripts blocks) which will make a request to the target application. This automatically appends session token to the request when the victim is logged in to the application at that time. The application will automatically perform whatever action the attacker requested.

The Target Page To execute this attack, the victim will access a specific page that looks like the page shown on Fig. 7.6

Fig. 7.7 A web page
which is vulnerable to
CSRF

The markup to create the page shown in Fig. 7.6 looks like the following:

```
<h1>Change your data</h1>
<form method='post'>
New value: <input type='password' name='newpassword'/>
<input type= 'submit' value= 'Change'>
</form>
```

This renders to page like this, hosted on a domain http://www.mydomain.com/change_your_data.html

When a logged in user hits the submit button, the browser knows it has to post the new password to greatdomainname.com. By convention, it will send the session ID along with the post request. When the user hits the submit button, the browser will send the new password to mydomain.com. In addition, the session ID will be send with post request.

The attacker will design a page such as the one shown in Fig. 7.7.

```
<h1>Free book!</h1>
<form method="post" action=http://www.mydomainame.com/change_data.
html>
    <input type="hidden" name="newpassword" value="h4ck3d!" /><br />
<br >
<input type="submit" value="Click here for your free book!" />
</form>
```

Notice that the attacker has instructed the form to submit to another domain then it is hosted on. The data in the attacker form will be directed to the attacker page. The user now logged in mydomain.com, and when he browses the attacker page. When the attacker clicks the button "click here" for your free book, this composes a request with a valid session ID, but the attacker has chosen his own password by using a hidden input field with the same name.

Database Security

Database is a collection of interrelated data and a set of programs to access the data. Database provides a convenient and efficient processing of data. A database management system is usually used to read, update and delete database components.

Database security refers to the collective measures used to protect and secure a database or database management software from illegitimate use and malicious threats and attacks. Data base security measures aims at protecting data from unauthorized disclosure, denial of service attacks, and unauthorized modification. There are several security measures to protect databases including:

- Security Policy
- Access control models
- Integrity protection
- Privacy problems
- Fault tolerance and recovery
- Auditing and intrusion detection

Access Control

Access control ensures that all direct accesses to object are authorized. It protects against accidental and malicious threats by regulating the read, write and execution of data and programs. There are several access control mechanisms to protect databases:

- Grant and Revoke
- Security through Views
- Stored Procedures
- Query modification
- Multilevel Security

Grant and Revoke

GRANT <privilege> ON <relation>
To <user>
[WITH GRANT OPTION]

- GRANT SELECT * ON *Registration* TO Matthews
- GRANT SELECT *, UPDATE(CLASS_NAME) ON *REGISTRATION* TO FARKAS
- GRANT SELECT(NAME) ON *Student* TO Brown

GRANT command applies to base relations as well as views.

Security through Views

This can be done by Assigning rights to access predefined views

CREATE VIEW Outstanding-Student
AS SELECT NAME, COURSE, GRADE

FROM Student
WHERE GRADE > B

The problem when using view for security is that views are difficult to maintain

Stored Procedures

In this type of countermeasures, we Assign rights to execute compiled programs using the following command

```
GRANT RUN ON <program> TO <user>
```

However, the problem with this kind of countermeasures is that programs may access resources for which the user who runs the program does not have permission

Query Modification

Query modification is done according to the privilege identified to each user. For example if the following privileges are given GRANT SELECT (NAME) ON *Student* TO Blue WHERE COURSE="CSCE 590" and Blue write the following query:

- SELECT *
  FROM *Student*
  Then the Modified query becomes

  - SELECT NAME
    FROM *Student*
    WHERE COURSE="CSCE 590"

Multilevel Security
Multilevel security is to present users at different security level, different versions of the database. The main problem of this approach is that different versions of the database need to be kept consistent and coherent without downward signaling.

Privacy Problems in Databases

Statistical database is a database which provides statistics on subsets of records. Statistics may be performed to compute SUM, MEAN, MEDIAN, COUNT, MAX AND MIN of records. Given those statistics, there are several types of data compromises in the database as follows:

Name	Age	Has diabetes
Alice	25	Y
Bob	34	Y
Frank	33	Y
Ivy	42	N
Jim	12	Y

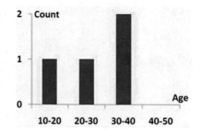

Fig. 7.8 How to compromise data in statistical database

	Total Income		Mr. White's Income
	Before the Move	After the Move	
Raw	$50,000,000	$49,000,000	$1,000,000
Privatized	$49,082,500	$48,734,148	$348,352

Fig. 7.9 How to compromise data in statistical database

- Exact compromise – a user is able to determine the exact value of a sensitive attribute of an individual
- Partial compromise – a user is able to obtain an estimator for a sensitive attribute with a bounded variance
- Positive compromise – determine an attribute has a particular value
- Negative compromise – determine an attribute does not have a particular value
- Relative compromise – determine the ranking of some confidential values

As an example Fig. 7.8 shows how the data is compromised in statistical databases, assume we have a database of medical records from a specific hospital, where 'Y' or 'N' denotes whether a person has diabetes or not. If the hospital intends to directly release a statistical report on the age distribution of diabetic patients in the database (see the histogram below), the privacies of individuals may be leaked. For example, if someone (often termed as an adversary) happens to know Alice is the only person who is between 20 and 30 years old in the hospital, he can confirm that Alice has diabetes.

The sources of this problem is due to adding or deleting records as shown in Fig. 7.9

Suppose that Mr. White is in a specific income database, let's assume the total income in his original neighborhood is $50 million. After he leaves, this Figure drops to $49 million. Therefore, one can infer that his true income is $1 million. To keep his income private, we have to ensure the query response is noisy enough to 'hide' this information

There are several methods to avoid data compromise, including:

- Query restriction
- Data perturbation/anonymization
- Output perturbation

Table 7.2 A database table with NoC as a private attribute

Name	Position	Age	NoC
George	Instructor	66	5
Heidi	Assistant professor	20	6
Holly	Associate professor	37	2
Leonard	Instructor	70	5
Massey	Instructor	33	3

Table 7.3 Query answered/Not answered based on QSOC

Query	Can be answered based on? (Yes, No)	Query result
Sum_NoC (position = Instructor)	Yes	13
Count (age < 60 & position = Instructor)	Yes	1
Sum_NoC (age > 60 & position = Instructor)	No	10

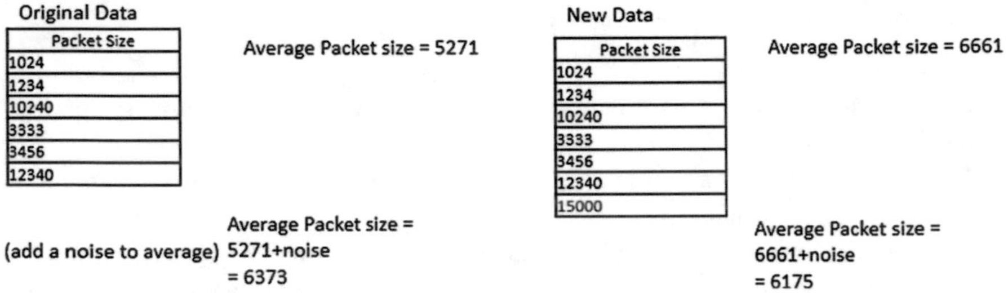

Fig. 7.10 Output perturbation example

Query restriction Based on Query Size A prevention mechanism that answers queries only when the size of the intersection of the query set and each previous query set is smaller than some parameter r. Consider Table 7.2 with r = 2.

Assume that NoC (number of children) is a private attribute. Based on **Query-set-overlap control (QSOC)** and **inference attacks** on statistical databases, if the queries in the Table 7.3 are to be executed, the first two queries can be answered without any privacy concerns. The last query cannot be answered since it will reveal the Number of Massey's Children.

Output Perturbation In this approach, the noise added to the output. For example in Fig. 7.10, without noise if the attacker knows the average packet size before the new packet is added, it is easy to Figure out the packet's size from the new average. With noise: One cannot infer whether the new packet is there.

Data Perturbation/Anonymization In this approach, the data itself is perturbed. For example, in the aggregation approach Fig. 7.11, the raw (individual) data is grouped into small aggregates before publication. The average value of the group

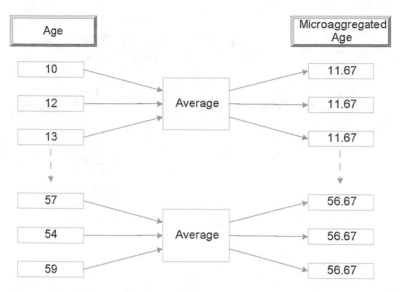

Fig. 7.11 Output perturbation example

replaces each value of the individual. Data with the most similarities are grouped together to maintain data accuracy. This approach helps to prevent disclosure of individual data.

SQL Injection Attacks

What is a SQL Injection Attack? Many web applications take user input from a form and this is often a user input that is used literally in the construction of a SQL query submitted to a database. For example: SELECT product data FROM table WHERE productname = 'user input product name'. A SQL injection attack involves placing SQL statements in the user input as shown in Figure. "Many, many sites have lost customer data in this way. SQL Injection attacks are often automated and many website owners may be blissfully unaware that their data could actively be at risk. These attacks can be detected and businesses should be taking basic and blanket steps to block attempted SQL Injection, as well as the other types of attacks we frequently see. Yahoo Voices was hacked in July 2012. The attack acquired 453,000 user email addresses and passwords. The perpetrators claimed to have used union-based SQL injection to break in. In addition, LinkedIn.com leaked 6.5 million user credentials in June. A class action lawsuit alleges that the attack was accomplished with SQL injection. SQL injection was documented as a security threat in 1998, but new incidents still occur every month. Making honest mistakes, developers fail to defend against this means of attack, and the security of online data is at risk for all of us because of it (Fig. 7.12).

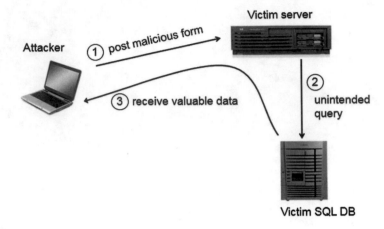

Fig. 7.12 Initiating SQL injection attack

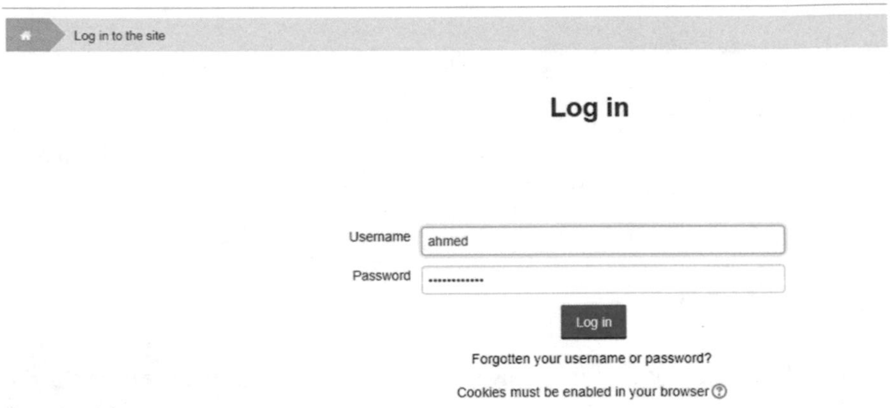

Log in

Username	ahmed
Password	••••••••••••

Log in

Forgotten your username or password?

Cookies must be enabled in your browser ⑦

Fig. 7.13 Typical login prompt

This is Input validation vulnerability as shown in Figs. 7.13, 7.14, 7.15, and 7.16. Unsansitized user input in SQL query to back- end database changes the meaning of the query.

Using SQL Injection to Steal Data

SQL Injection is a web based attack used by hackers to steal sensitive information from organizations through web applications. It is one of the most common application layer attacks used today. We will focus on three operators that are used to execute SQL injection attacks

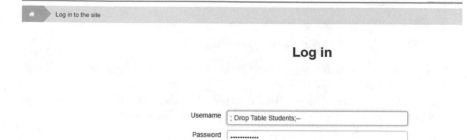

Fig. 7.14 Malicious user input

Fig. 7.15 Back end query
formulation

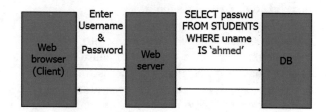

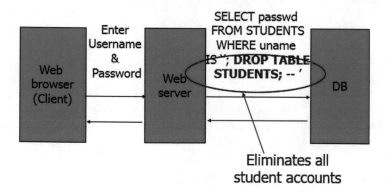

Fig. 7.16 Unsanitized user input in SQL query execution

A. Steal Data (Example) using OR operator

- User gives username ' OR 1=1 –
- Web server executes query
  SELECT * FROM UserTable WHERE
- username='' OR 1=1 -- …);
- Everything after -- is ignored!

- Now all records match the query
- This returns the entire database

B. Steal Data (Example) using LIKE operator

- To authenticate logins, server runs the following SQL command against the user database:
  SELECT * from UserT WHERE user='name' AND pwd='passwd'
- User enters' OR WHERE pwd LIKE '% as both name and passwd
- Server executes
      SELECT * from UserT WHERE user=" OR WHERE pwd LIKE '%'
  AND pwd=" OR WHERE pwd LIKE '%'
- Logs in with the credentials of the first person in the database (typically, administrator!)

C. Another Malicious User Input to Drop Tables

The attacker might use malicious database command to drop the table as follows:
Query'; DROP TABLE prodinfo; --

This Results in the following SQL: SELECT prodinfo FROM prodtable WHERE prodname = 'blah'; DROP TABLE prodinfo; --'

Notice that that last comment (--) consumes the final quote which causes the entire database to be deleted. The success of this attack depends on knowledge of table name. This is sometimes exposed to the user in debug code called during a database error. To overcome this type of attacks there is a need to use non-obvious table names, and never expose them to user.

Mitigation of SQL Injection Attacks

The main mitigation of SQL injection attacks is to sanitize the input. Overall, web application developers often simply do not think about "surprise inputs", but security people do (including the bad guys). Sanitizing user inputs to insure that they do not contain dangerous codes, whether to the SQL server or to HTML itself. One's first idea is to strip out "bad stuff", such as quotes or semicolons or escapes

Ex. An email address only contains "A B C D E F G H I J K L M N O P Q R S T U
     V W Y Z 0 1 2 3 4 5 6 7 8 9 @ . -_"

Skills Section

In this section we will introduce a complete example on SQL injection. The following web application accepts the email address, and password then submits them to a PHP file named index.php.

Consider a simple PhP web application with a login form. The code for the HTML form is shown below.

In this section we will focus on SQL injection Attacks and how they are initiated

```
>form action='index.php' method="post">
>input type="email" name="email" required="required"/>
>input type="password" name="password"/>
>input type="checkbox" name="remember_me" value="Remember me"/>
>input type="submit" value="Submit"/>
</form>
```

We will illustrate SQL injection attack using SQLfiddle. The following schema will be created as part of this example:

```
SQL Fiddle
1 CREATE TABLE `users` (
2    `id` INT NOT NULL AUTO_INCREMENT,
3    `email` VARCHAR(45) NULL,
4    `password` VARCHAR(45) NULL,
5    PRIMARY KEY (`id`));
6
7
8 insert into users (email,password) values ('m@m.com',md5('abc'));
```

Once the schema is created we will insert a DB record as shown below

id	email	password
1	m@m.com	900150983cd24fb0d6963f7d28e17f72

Let's suppose an attacker use the following input in the email address field.
xxx@xxx.xxx' OR 1 = 1 LIMIT 1 -- ']
The generated dynamic statement will be as follows.
SELECT * FROM users WHERE email = 'xxx@xxx.xxx' OR 1 = 1 LIMIT 1 -- '] AND password = md5('1234');

```
1 SELECT * FROM users
2 WHERE email = 'xxx@xxx.xxx'
3 OR 1 = 1 LIMIT 1 -- ' ] AND password = md5('1234');
```

The statement xxx@xxx.xxx ends with a single quote which completes the string quote. The OR 1 = 1 LIMIT 1 is a condition that will always be true and limits the returned results to only one record. The statement -- ' AND ... is a SQL comment that eliminates the password part. The query will retrieve the following record.

id	email	password
1	m@m.com	900150983cd24fb0d6963f7d28e17f72

SQL Inject a Web Application

A simple web application at http://www.techpanda.org/ that is vulnerable to SQL will be used to demonstrate injection attacks for demonstration purposes only.

www.techpanda.org

Login | Personal Contacts Manager v1.0

Email*

xxx@xxx.xxx

Password*

....

Remember me

Submit

The generated SQL statement will be as follows:
SELECT * FROM users WHERE email = 'xxx@xxx.xxx' AND password = md5('xxx') OR 1 = 1 --]');

Other Forms of SQL Injection Attacks

SQL Injections can do more harm than just by passing the login algorithms. Some of the attacks include

Deleting data
Updating data
Inserting data

In this practical scenario, we are going to use Havij Advanced SQL Injection program to scan a website for vulnerabilities.

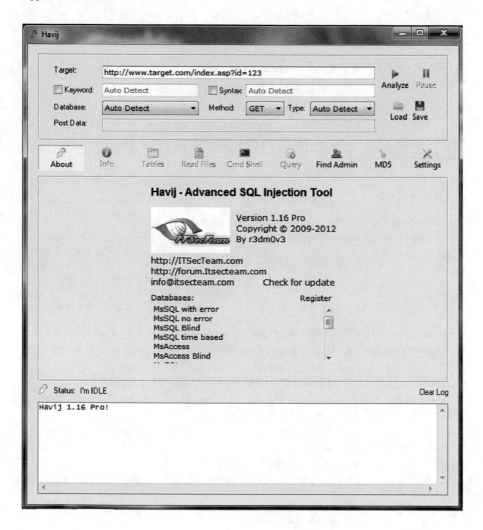

Applications Section

Applications for Encryption Techniques for Network Security

- You are expected to implement specific web application and then test it against web attacks such as SQL injection and XSS and CSRF

Applications of Website Vulnerability Scanners

- You are expected to search for scanners that can discover website vulnerabilities.
- Install and run those scanners on particular websites

Applications for Detecting Network Attacks Using Traffic Analysis

- You are expected to select one of the known vulnerable websites.
  - Check if the has any SQL injection vulnerabilities
  - Classify those vulnerabilities into insert update and delete vulnerabilities

Questions

- In the scope of web security, describe the differences between XSS and CSRF
- Following is a script that has been written for a login page in a specific web application txtUserId = getRequestString("UserId");
- txtSQL = "SELECT * FROM Users WHERE UserId = " + txtUserId;
- Assume that there is an SQL vulnerability in this application. Write down a query that return ALL rows from the "Users" table
- If you have the following part of a web application uName = getRequestString("username");
  uPass = getRequestString("userpassword");
  sql = 'SELECT * FROM Users WHERE Name ="'+ uName + '"AND Pass ="'+ uPass + '"'

User Name:
[   ]

Password:
[   ]

- Describe two vulnerabilities that can lead to XSS and SQL injection attacks
- Describe two techniques to mitigate SQL injection attacks
- What is the main purpose of query modification when an unauthorized query is written.

References

Application for testing and sharing. SQL queries. http://sqlfiddle.com
Cross site Scripting attacks. http://deadlytechnology.com/web-development/xss/
SQL Injection. https://www.w3schools.com/sql/sql_injection.asp
SQL Injection practical Example. https://www.guru99.com/learn-sql-injection-with-practical-example.html
SQL Injection practical Example. http://www.techpanda.org/

Chapter 8
Mobile and Wireless Security: Lesson Plans

Competency: Learn major aspects of user/software security issues in mobile/smart devices.
Activities/Indicators
• Study reading material provided by instructor related to major aspects of user/software security issues in mobile/smart devices.
• Complete successfully an assessment provided by instructor related to competency content.
• For mastering levels, more than 80% of assessment grades should be earned in no more than three trials.
• Assessment questions can be pulled from the end of the chapter questions or any relevant material.
Competency: Learn major aspects of operating system/platform security issues in mobile/smart devices.
Activities/Indicators
• Study reading material provided by instructor related to major aspects of operating system/platform security issues in mobile/smart devices.
• Complete successfully an assessment provided by instructor related to competency content.
• For mastering levels, more than 80 % of assessment grades should be earned in no more than three trials.
• Assessment questions can be pulled from the end of the chapter questions or any relevant material.
Competency: Learn major aspects of network/environment security issues in mobile/smart devices.
Activities/Indicators
• Study reading material provided by instructor related to major aspects of network/environment security issues in mobile/smart devices.
• Complete successfully an assessment provided by instructor related to competency content.
• For mastering levels, more than 80% of assessment grades should be earned in no more than three trials.
• Assessment questions can be pulled from the end of the chapter questions or any relevant material.

(continued)

© Springer International Publishing AG 2018   159
I. Alsmadi et al., *Practical Information Security*,
https://doi.org/10.1007/978-3-319-72119-4_8

Competency: Learn major aspects of network/environment security issues in wireless networks

Activities/Indicators

* Study reading material provided by instructor related to major aspects of network/environment security issues in wireless networks
* Complete successfully an assessment provided by instructor related to competency content.
* For mastering levels, more than 80% of assessment grades should be earned in no more than three trials.
* Assessment questions can be pulled from the end of the chapter questions or any relevant material.

Competency: Learn major aspects of access controls in mobile and wireless platforms.

Activities/Indicators

* Study reading material provided by instructor related to major aspects of access controls in mobile and wireless platforms.
* Complete successfully an assessment provided by instructor related to competency content.
* For mastering levels, more than 80% of assessment grades should be earned in no more than three trials.
* Assessment questions can be pulled from the end of the chapter questions or any relevant material.

Competency: Learn how to evaluate user/software security issues in mobile/smart devices.

Activities/Indicators

* Use one or more open source or free tools that allow users to evaluate user/software security issues in mobile/smart devices.
* For mastery, student is expected to try more than one operating system.
* For a mastering level, student should show what tool they selected, how they installed the tool and/or present a video to demonstrate the different commands they have used.

Competency: Learn how to evaluate operating system/platform security issues in mobile/smart devices.

Activities/Indicators

* Use one or more open source or free tools that allow users to evaluate operating system/platform security issues in mobile/smart devices.
* For mastery, student is expected to try more than one DBMS.
* For a mastering level, student should show what tool they selected, how they installed the tool and/or present a video to demonstrate the different commands they have used.

Competency: Learn how to evaluate access controls in mobile and wireless platforms.

Activities/Indicators

* Use one or more open source or free tools that allow users evaluate access controls in mobile and wireless platforms.
* For mastery, student is expected to try more than one website or web server.
* For a mastering level, student should show what tool they selected, how they installed the tool and/or present a video to demonstrate the different commands they have used.

Competency: Learn how to evaluate network/environment security issues in mobile/smart devices.

Activities/Indicators

* Use one or more open source or free tools that allow users to evaluate network/environment security issues in mobile/smart devices.
* For a mastering level, student is expected to show more than one RBAC example or application.
* For a mastering level, student should show what tool they selected, how they installed the tool and present a video to demonstrate the different commands they have used.

Competency: Learn how to evaluate network/environment security issues in wireless networks

Activities/Indicators

* Use one or more open source or free tools that allow to evaluate network/environment security issues in wireless networks. Instructor can make such judgment.
* For a mastering level, student is expected to show more than one OBAC example or application.
* For a mastering level, student should show what tool they selected, how they installed the tool and present a video to demonstrate the different commands they have used.

Overview

Mobile and wireless usage by humans continue to grow rapidly in the last few years. Number of subscribers to mobile service provides exceed the number of human population (As in many cases, users carry more than one mobile or phone). Many consumer reports showed that in the recent years the number of smart devices sold exceeds the number of desktops and laptops. Many users are shifting their usage in many aspects from desktops and laptops to smart devices. Mobile internet usage, e-commerce, Online Social Networks (OSNs), etc. are all showing mobile platforms take-over from desktops and laptops.

The age of using mobile phones went down where recently reports showed that the population of kids below the age of 10 using mobile phones is also growing. Additionally, the level of usage for mobile phones is also growing. Most users now user for mobile phones smart devices that can have almost the same capabilities of a tablet or even a laptop on a mobile operating system. In order to distinguish wireless service from mobile services, wireless services largely refer to (WiFi) services especially for Internet-based services. Those can now be used for smart phones, desktops, laptops, tablets, TVs, etc. There are also some other types of wireless services, in addition to WiFi that we will cover briefly in this chapter.

In this chapter, our goal is to describe security issues related to mobile and wireless platforms. We will divide them first based on mobile or wireless services. For mobile-based services, we will classify the sections based on 3 categories: Mobile user/software related security issues, mobile operating system or platform security issues and finally mobile network or environment security issues. In the wireless section, we will focus on wireless major platforms and related security issues. We will also cover a separate section for access controls in mobile and wireless platforms.

Knowledge Sections

In the knowledge section, different mobile and wireless security aspects will be discussed. As most of the current mobile phones fall within the categories of smart phones or devices, we will use the term smart devices or phones from now and afterwards. For smart devices, we will divide security issues in three sections, security issues in software applications (excluding platforms or operating systems). The second part will cover security issues in mobile operating systems. The third section will cover security issues on smart devices networks in general. This third category focuses on hardware and network components. In wireless section, we will cover one main section in security issues in different wireless networks. Last part will cover access control issues in smart and wireless platforms in general.

Users/Software Security in Mobile/Smart Devices

Recent reports indicate that the nature and the volume of security attacks on mobile/smart devices are growing rapidly. Attackers continuously screen those devices or their networks for possible vulnerabilities. They also continuously improve the methods and techniques of targeting those devices. The source of vulnerability can be one or a combination of more than one. Vulnerabilities may exist in the device hardware, network, operating system, application and finally the users themselves and their "poor" security practices. There are different types of security controls that are implemented and enabled for users to set and control. Users may be reluctant to spend the time and effort to understand and work with those security controls. In some other cases, some security controls may impact users' preferences or convenient options (e.g. location-based services). Users may completely opt-out using such security controls once they feel that they can't use some services easily or conveniently.

Application Code Signing (Vetting)

The goal of signing an application is first to identify who made/signed the application and second to check/compare any application against this "original" identity. Mobile application stores require that applications should be first signed before it can be posted in the store. Mobile operating systems do not also allow unsigned applications to operate. However, Android may allow self-signed applications to operate. Apple Xcode developers' IDE allows developers to sign their developed code with their unique identity (that they created through Xcode). Eclipse IDE can provide keys for Android applications. The digital identity contains a pair of private/public keys (PKI) and a digital certificate that can uniquely, based on hardware configuration, identify certificate holder/owner. The certificate authority (CA) to issue those certificates for Apple application store is Apple. Xcode or Keychain access can be used to manage identity management. The operating system can check if any changes occurred to the signed application (e.g. due to a malware). Hashes are created for all resources in the application. Any resource change in the application will lead to a change in the application hash set. The application hash-set is encrypted with a private key. The public key is distributed with application package.

Bring your Own Device (BYOD) and Security Issues

Companies struggle to define best policies to propose and enforce on how employees should use smart devices in the work environment. There are several factors that should be taken into consideration and balanced when making such policies. Main criteria to consider include: Security and privacy issues, convenience and realistic policies (i.e. to be able to apply and enforce such policies) and finally the level of data sensitivity that exist in work environments. We select the following issues to discuss in this scope:

- Company or employee smart devices: Should companies provide smart devices to employees to use them in the work environment and for the work functions or should the company allow or ask users to use their personal smart devices?
- Policies to make to respond to this question have many dimensions: The first one can be the cost dimension, to buy those smart devices for all/selected employees and to maintain those devices in cases of problems. The second issue is whether to allow employees to download company applications and data on their personal devices or allow them to use the company smart devices for some personal applications or usage.
- Policies' enforcement: From practical considerations, it's uneasy to enforce roles or policies for employees on what they can bring, use or install on the smart devices they own or they have.

General security related policies for any company must be aligned with company vision and mission and its business functions. In some cases, users will not be allowed to use their own devices to access company network, domain or resources. They need to completely separate their personal from their business related computer usage. In those types of companies or businesses, all smart devices should be assumed un-trusted. They can be allowed to access company resources with only accounts that have limited access controls. As an alternative, employees can use smart devices provided and fully secured and tested by the company and only for company related usage. However, when business nature does not involve highly classified data or activities, policies should exist to allow, but regulate employees when using their personal or mobile devices.

Theft and Loss/Unauthorized Access Issues

Mobile devices are easy to carry, in comparison with desktops, laptops and tablets, and they are also easy to lose, forget, or steal. While some devices can be monetary expensive, however, for most users, the actual loss will be the amount of value and information that they may lose when they lose the phone. Additionally, serious security and privacy concerns arise especially as users are used their phones to store all their personal documents, pictures and videos. They also have direct access through the phone to their bank accounts, personal emails, Online Social Networks (OSNs) and favorite e-commerce websites. A user who can find/use this phone can easily claim their identity in many websites. One problem is that all those types of websites trigger user Personal Identification Number (PIN) sent to their mobile once they try to access the website from a *new* location. However, there is no logical way for those websites to know that the current mobile holder is not the actual user. Manufactures of smart phones include a feature to allow users to create a master

PIN to lock or unlock the main phone. For example, in Apple IPhones ™ users will not be able to unlock the phone without the PIN except if they decided to follow the steps (see skills section) to erase all phone data and return it to manufacture, empty, state. If users create previous backup points (e.g. using Apple iTunes ™), they can restore the phone from the backup restoration points. iOS includes also a feature to wipe phone data after 10 incorrect tries of entering the password.

The second and major phone lock option Apple IPhones ™ offer is called (Cloud lock). In this case, the phone will not open nor restore unless unlocked through original user account. In this case, the phone can be only usable for parts and not as a whole phone set. Students can search the Internet for offered procedures to unlock (Apple IPhones ™ cloud locks).

Mobile Security/Malware Threats

Malware threats such as viruses, worms, spywares, Trojan horses etc. typically target desktop and web platforms. Malwares in generally can be distinguished based on several factors:

- Access or intrusion method: Different malwares have their different ways of getting to victim machines. They can trick users to install them through masquerading as genuine applications. They may also fake their identities to look like genuine applications or embed themselves in innocent files. Users can be also tricked to install those malwares through different social engineering methods. For smart devices in particular, downloading some free applications from untrusted application stores seem also to be a popular access method for those malwares.
- Propagation: Different malwares have their different propagation or expansion methods. The worst and most serious in terms of propagation are worms and spams as their main goal is to expand to the largest number of possible machines, smart devices or users. This expansion can be triggered through accounts (i.e. victim friends' list) in their accounts or email applications. They can be also triggered through certain networking schemes (i.e. expand gradually through local networks of victims). They do not require, usually, users' intervention to make the propagation trigger.
- Payload: Malwares are malicious applications developed usually to cause harm to victim machines. Such harm or payload can vary from one malware to another. Many complex malwares can have many types or categories of payloads. For example, the main payload of a spyware is to steel users' private information, financial data and possibly to commit identity theft. The main goal of an adware is to keep post unsolicited ads and links to users. The goal of mobile auto-rooters is to gain "root" level access privileges. This may usually lead to other types of payloads.

- Hiding methods: Recent advanced and complex malwares employ methods to hide, avoid detection, and avoid analysis or reverse engineering analysis. They typically will use some encryption or obfuscation method to defeat one or more of those previously mentioned counter attack methods. They may also hide or delay their payload to avoid detection. They may also act or behave differently in different time, environments, etc. to avoid or complicate detection methods.

Usage of Smart devices as a platform is growing rapidly in number of users as well as applications. While large scale mobile malware infections are insignificant yet, this is expected to change in future. Both mobile manufacturers and users should have the vision to expect that this may happen and should be prepared for it. Followings are examples of recent mobile malwares:

1. The Pegasus spyware: Spywares try to steal different types of information based on the nature of the attacks (e.g. identity theft, private data, pictures, listen to your audios, screen, etc.). One of the most recent mobile malwares is Pegasus spyware. The malware is traced to a malware company in Israel (NSO Group) owned by a private US equity. The malware was discovered and Apple iOS released 9.3.5 to include required security updates within 10 days from the first time a human right defender received a message related to the malware. The message was forwarded and analyzed by Lookout security company.

Pegasus focuses on utilizing tools that come with most smart devices (e.g. WiFi, Camera, Microphones) to spy on users' videos, audios, messages, accounts, etc. The malware uses also encryption to avoid being easily detected. The malware analysis showed that it exploits 3 types of zero-day vulnerability in Apple iOS ™:

- CVE-2016-4655 (https://web.nvd.nist.gov/view/vuln/detail?vulnId=CVE-2016-4655): Kernel Information leak in Kernel. Apple iOS ™ (before 9.3.5) allows attackers to obtain sensitive information from memory. A kernel base mapping vulnerability exists that leaks information allowing the attacker to calculate the kernel's location in memory.
- CVE-2016-4656 (https://web.nvd.nist.gov/view/vuln/detail?vulnId=CVE-2016-4656): Kernel, Apple iOS ™ (before 9.3.5), Memory corruption leads to silently jailbreak the device and install surveillance software.
- CVE-2016-4657 (https://web.nvd.nist.gov/view/vuln/detail?vulnId=CVE-2016-4657): Memory corruption to execute arbitrary code or cause a DoS in Apple iOS ™ (before 9.3.5). The vulnerability is exploited in the Safari WebKit that allows the attacker to compromise the device when the user clicks on a link.

Initially, the malware get started through a phishing message users can receive in their phones. Once the user clicks or activates the message, it opens a web browser which eventually activates the malware payload silently without any notice from the user. The spyware payload includes enabling the attacker to expose different user applications such as: Messaging applications, emails, WhatsApp, FaceTime,etc.

2. **Shedun, Hummingbad, Hummer, Shuanet, ShiftyBug, Kemoge, GhostPush, Right_core, and Gooligan**

All those names of malwares in Android platform are believed to be connected with each other. Malware investigators showed also that the majority of the source code for those different malwares is similar. Early versions of this malware appeared in 2015. They may represent aliases for the same original malware with some slight variations in: signature, payload, access method, functionalities, etc. The malware infects devices through the installation of some applications that include the malware. The add-on adware is added to normal and popular applications that users often download and use (e.g. Twitter, WhatsUp, Facebook, Google, etc.). The new packaged applications are then made for public download in third party application stores. One of the payloads for the malware is exposing Google accounts for the victim user. The malware series have also the ability to tamper with Android accessibility services. KitKat and Jelly Bean are the most widely impacted Android systems by this type of malware. Malware payload can be also creating a backdoor to allow attacker to access victim mobile remotely. This is achieved by "auto-rooting" to elevate attacker privilege to a "root" privilege through exploiting some of the popular Android exploits (e.g. ExynosAbuse, Memexploit and Framaroot).

3. **Dendroid**: A malware that targeted also Android platform. It was discovered in 2014 as a malware to allow attackers to gain remote control to victim mobile phones. Google Bouncer anti-malware that is deployed in Google Play was unable to detect this malware. It is believed that this malware uses remote control methods similar to those of earlier Zeus and SpyEye malwares. The malware was developed to be packaged with other legitimate mobile applications. As a typical remote control malware, this malware can perform several types of actions on remote victims' mobile phones.

4. **ViperRAT**: A recent Advanced Persistent Threat (APT) malware that targeted Israel defense force. APT refers to the list of persistent malware threats with advanced and stealthy natures of hacking or malware activities. The term is used to label different categories or types of malwares in different platforms, not only in mobile devices. An APT can also include different malwares in different platforms. ViperRAT main goal was to spy on mobile victims, steal data and monitor audio and video recordings, contacts and photos. Some versions of ViperRAT are believed to be deployed to victim machines through Social engineering methods using fake social networking profiles. Payload of the malware is installed as a Viber application update or as a general software update.

5. **Pretender Applications:** There are many incidents where fake applications exist in the mobile market application stores. Those fake applications use the name, logo, etc. of popular legitimate applications that users often use or download. This is an old and classical trick for some malwares for an easy deployment method. Users who download software applications from unknown market stores may not notice any difference between original and fake application. Even if

some slight differences exist, they may think its just a new or enhanced version of the version they are familiar with.

6. **Triada:** This is a backdoor malware that targets Android systems. It also employed IP address spoofing in loaded web pages.

7. **Hiddad:** Another backdoor malware that targets Android systems. Similar to many other Android malwares, it repackages legitimate applications, include itself and then deploy using the same names of the legitimate applications in third-party application stores.

8. **RuMMS:** An android-based malware family targeted users in Russia via SMS Phishing (Smishing). SMS phishing messages that contain malicious links are sent to victims mobile phones. The main payload in the malware is related to identity theft and stealing privacy and finance related information.

9. **Brain Test:** This Android application was in Google Play store for a while masquerading as a legitimate application. The application, believed to be primary for advertisement payloads, is masquerading as an IQ testing application. Google has "Google Bouncer" anti malware scanning program to scan applications in Google Play store. However, the Bouncer fails to detect the malware in this program for some time. In an advanced case, once the malware was detected, it was introduced again with an obfuscation mechanism to distort the ability to detect this malware as the earlier version.

10. **XcodeGhost:** An iOS malware that infects Xcode compiler and Integrated Development Environment, (versions 6.1 and 6.4). Xcode is used by iOS to develop mobile applications. The malware was first seen in the Chinese website and search engine Baidu. It is believed that more than 50 mobile applications that were developed and deployed were infected by this malware. Some examples of the popular applications that were infected include: WeChat, Didi Chuxing, Railway 12,306, WinZip and Tonghuashun. An author claimed creating this malware and posted its code openly in Github. For users it was hard to judge if this is a malware as it came through legitimate applications and store. Most payloads discovered related to this malware are spyware related activities such as spying on users' private information, users' credentials, etc.

Installing Mobile Applications from Unknown or Un-trusted Stores

For the different mobile platforms, certain application stores exist to allow users to search for and download applications from. Security and trust of software vendors is provided or guaranteed by the third-part; the app. Store host. Users may search for other unknown or un-trusted application store websites to find free versions of applications. Such unknown websites can be markets for malwares of different types. Users should be very careful and make sure to avoid installing software applications from such websites or stores. Particularly in the Android platform, many examples of malwares spread through files downloaded from un-trusted app-stores.

Mobile Anti-malware Systems

The market of anti-malware programs for smart devices is growing. Figure 8.1 below shows examples of those anti-malware programs (from Google play applications market). Few anti-malwares dedicated for smart devices exist in the industry. Many unique characteristics exist in mobile devices make the classical desktop/laptop antimalware systems incapable of dealing with malware challenges in smart devices' environments.

The first category in those anti-malware systems include the traditional anti-malware systems companies known before the evolution of mobile platforms or operating systems. Those include: Kaspersky, Bitdefender, Avast, Avira, McAfee, ESET, and Norton. On the other hand, there are anti-malware vendors that are (mobil-only). Lookout is the best example in this category which is taking a significant share of mobile anti-malware vendors. Users can expect such products to be lighter as they are dedicated for mobile platforms. Other mobile-only anti-malwares include: TrustGo, AegisLab and SuperSecurity.

By default, mobiles do not come with anti-malwares pre-installed. Users may not install, on their own, anti-malware systems for cost purposes or as they may slow down their systems or consume the battery. Many users may not be aware of the seriousness of malwares and hence may not be willing to install anti-malwares on their own.

In addition to antimalware systems, mobiles may also need firewalls/Intrusion Detection/Protection Systems (IDS/IPS). While those applications are avoided in mobiles due to their possible overheads and resources' consumption, yet with time, lightweight forms of those security controls will possibly be necessary. Anti-malware applications for example may not act like a firewall to monitor and block all in/out ports that are not used by legitimate user applications/traffic.

Fig. 8.1 A sample of Mobile anti-malware apps

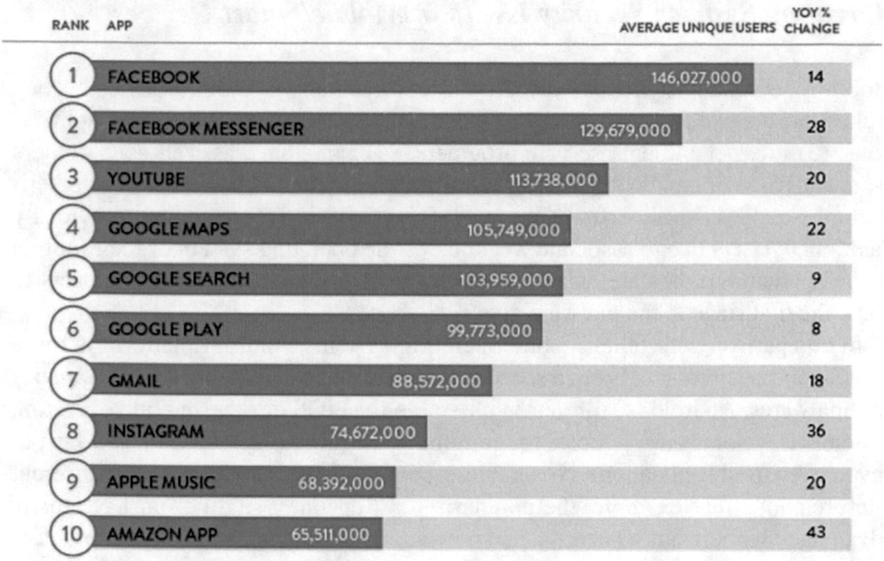

RANK	APP	AVERAGE UNIQUE USERS	YOY % CHANGE
1	FACEBOOK	146,027,000	14
2	FACEBOOK MESSENGER	129,679,000	28
3	YOUTUBE	113,738,000	20
4	GOOGLE MAPS	105,749,000	22
5	GOOGLE SEARCH	103,959,000	9
6	GOOGLE PLAY	99,773,000	8
7	GMAIL	88,572,000	18
8	INSTAGRAM	74,672,000	36
9	APPLE MUSIC	68,392,000	20
10	AMAZON APP	65,511,000	43

Fig. 8.2 Top Smart phones apps in 2016 (http://www.nielsen.com)

Online Social Networks (OSNs)

Online Social Networks (OSNs) represent a major category of applications that users heavily used in mobiles. Typically, mobile applications such as: Facebook, Instagram, etc. (Fig. 8.2).

Such heavy usage of OSN applications make them focused targets for many types of malwares. Many malwares find it easy to pass their links through OSNs. This is since one of the main goals for any malware is to spread to a large number of possible victims within a short amount of possible time; No websites or applications can better serve this goal similar to OSNs. In some cases, attackers may trick some users in OSNs to add them to their friends' list. This may give the attackers the ability to post some links in the user profile or command with the user friends. For worms, this can be the perfect large scale and high speed propagation method. OSNs enable users to change/edit privacy or security settings. Users can typically decide to limit friends' visibilities/interactions with their profiles. They can else prevent friends or other users to post on their behalf, pages, walls, etc. However, in reality, few users may want to spend the time to frequently screen their privacy and security settings and verify their choices. They may lack the technical knowledge to make the right decisions.

Operating Systems Security Issues in Mobile/Smart Devices

Mobile operating systems are the ground applications on mobile devices. They orchestrate control and management activities between (1) users, (2) their applications, (3) Internet and data service providers, (4) app-stores, as well as (5) others users and (6) their mobile devices. As a result, security in mobile devices can inherit strengths and weaknesses from any category of those five categories previously mentioned. Users like to associate weaknesses the operating system of their mobile or the applications they are using. However, most attacks start by tricking users to make the first stage in the malware attack.

In comparison with other mobile operating systems, Android platform is more popular and open source. From a security perspective and based on several statistics and malwares, Android is still more vulnerable than iOS or other mobile operating systems. Developers use popular programming languages such as C++ and Java to develop Android applications. Nonetheless, such statistics can always have different interpretations. For example, the popularity and openness of Android have many advantages while from a security perspective, it is expected to be more targeted. Such exposure can bring more vulnerabilities and simultaneously more fixes to those vulnerabilities. Android allows installing applications from third party sources. While this has several advantages related to Android openness and flexibility, this seems to be one of the most serious sources of security vulnerabilities for Androids. While users should be aware of such issues and be careful of making decisions to install such applications from third parties, Android should find methods to make sure that malwares from those un-trusted websites can be quarantined. Mobile platforms may need to enforce security policies to third-party application providers. Users will then only be allowed to install applications from those providers once those applications pass security policies.

Apple showed in their website security architecture of iOS ™ (Fig. 8.3). The layered security architecture is designed to defend against different threats. Hardware and firmware security layers are designed to protect against malwares while high level OS layers are designed to provide a reliable access control architecture.

Apple iOS security.components can be summarized as:

- Boot-up process and secure boot chain.
- Software updates and Secure Enclave
- System Software Authorization
- Encryptions and hardware security features
- Encryptions and file data protections
- Pass-codes, passwords, etc.
- Application security: This includes: Application code signing, runtime process security, etc.
- Network security features (e.g. TLS and VPN).

Unlike iOS, Android does not enforce OS level access control on its applications to access mobile resources. Android developers can add permissions to their applications using certain programming tags.

Fig. 8.3 Apple iOS high
level security architecture
(source: Apple)

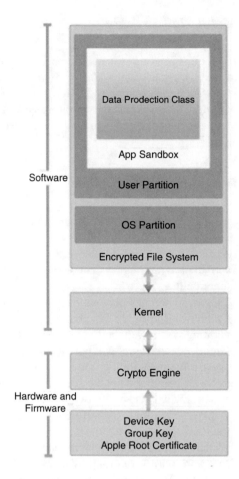

Sandboxing

Sandboxing is an environment where codes in the sandbox are isolated from the main environment. In the Sandbox each application has access to its own files, preferences, and network resources while operating system controls communication outsides the Sandboxing environment. In iOS, all third-party apps are "sandboxed,". They are restricted from accessing files stored by other applications and from making changes to the device.

In Android, User and Group Identifications (UID/GID) decides the permission model that provides data isolation. Two access control models are used to enforce data confidentiality: 1. Discretionary Access Control (DAC) that allows only the device owner to access their own files, and 2: Mandatory Access Control (MAC) that constrains the ability to access or perform certain operations on specific objects/targets.

Mobile Jail-Breaking

In Mobile operating systems it is unconventional to allow users to have "root" access to the operating system. This "privilege elevation to root" is called Jail-breaking. Users are not recommended to try to perform Jail-breaking, using certain methods or tools. This may open the operating system for several possible vulnerabilities and break the pre-designed operating system security architecture.

Operating System Updates

Operating systems offer frequent updates through the application store. Those updates can be triggered by new or modified features. It can be also triggered by security fixes for newly discovered vulnerabilities. It's very important for users to keep their operating systems up-to-date to avoid some instances of zero-day vulnerabilities.

Users sometimes may not be able to download some security updates. This can be either caused by the fact that the phone is old and is not supported anymore by operating system vendor. It is also possible that, due to size limitations, or operating system versions, users will not be able to download those security updates.

Security vulnerabilities may come from some software applications that the user has on the phone and not the operating system. Those application vendors may have a much slower cycle of discovering their vulnerabilities and making new releases to patch vulnerabilities. Even if they have them, users may not notice the need to do those necessary security updates.

There are many challenges in relation to security updates. First, the time to detect a malware may vary. Some malwares take few days/weeks/months before being discovered. Within such period, many mobiles can be infected. In the second stage, security researchers need to make their own investigations to first understand the malware, how it accesses victims' machines and what is the nature of the payload. This may help them eventually understand mobile software or operating system vulnerabilities that may seduce such malwares. In the last stage, they need to develop updates/fixes for the mobile applications or operating systems. This long cycle shows the complexity of continuously developing operating systems' updates to respond to malware breakouts.

Prepare for Phone Physical Theft

Statistics show that physical thefts of mobile phones continue to be a serious security related problem. When attackers have a physical access to mobiles, they can perform all kinds of harms and payloads. Users can protect against such security threats by encrypting their devices. For example, IPhones use iCloud locks to prevent phones from being accessed without successful login to owner registered iCloud account.

App Transport Security (ATS) is proposed recently by Apple to enforce all applications' traffic to go through encrypted HTTPS connections.

Hardware/Network Security Issues in Mobile/Smart Devices

In addition to vulnerabilities that may come from mobile operating system or installed applications, the mobile hardware and the network (i.e. data and voice providers) can be also sources of vulnerabilities that can be exposed by hackers. In this section, we will describe examples of security issues in the hardware and the networks of mobile phones.

Apple IPhones listed the followings as device/hardware security mechanisms or tools:

- Passcode protection
- iOS pairing model
- Configuration enforcement
- Mobile device management (MDM)
- Device restrictions
- Remote wipe
- "Find My iPhone" and "Activation Lock"

A Secure Phone Booting Process

In operating systems, the booting process includes the sequence of activities that start as soon as the user starts or restarts the system (e.g. the mobile device) till the system is ready for user normal usage. A secure phone booting process should include only the steps that are specified by the operating system without any possible tampering from an attacker or a malware. The phone will stay locked until all steps in the booting process are properly verified.

For example, recently Google implemented a "verified boot" process in Android to improve device security. The process uses cryptographic integrity checking to verify that device is not tampered with. If a tamper is detected, the device may start with very limited resources (more like a safe mode), or may not start (Fig. 8.4). The process includes also techniques for "automatic error corrections" that try to correct some errors without any user involvement. Verified boot process may also make it hard to jail-break a mobile operating system.

In iOS, when the low level boot-loader finishes its tasks, it verifies and runs, iBoot, which in turn verifies and runs the iOS kernel.

Fig. 8.4 Google verified boot warning messages

AES Crypto-Engine

Advanced Encryption Standard (AES) algorithm is a data scrambling system adopted by US government in 2001 and is widely accepted as *unbreakable*. A truly random 256-bit AES key will be impossible to break, given the current and near future technologies and techniques. In Apple, the encryption sets between the flash storage and the main memory. Two non-erasable device keys: (Group ID, GID and User ID, UID). They are built in the device hardware with 256-bit AES keys. Each key is burnt into the hardware and hence it can't be tampered by either the hardware or the firmware. This built-in encryption can also provide a fast way to perform remote system wipe. The encryption key is protected by the user's "PIN" a code that must be entered by the mobile user before the device can be used. Unlike iOS, Android has no key built or burned in the hardware.

iPhone can be programmed to wipe itself if the wrong PIN is provided for more than consecutive 10 times. Trying all four-digit PINs will require no more than 800 s. With a six-digit PIN, the maximum time required would be 22 h.

Access Control Models

In the scope of access controls in wireless networks, it is important to differentiate between Authentication, Authorization or Access control and Accountability (AAA). In Internet WiFi home-computing, you get one account which integrates the three in one account (with a user name and password). A typical home that has many desktops, laptops, tablets, or smart phones, will allow all those users to access home router using the same single account. Ideally, you want to give them all the same Authentication/ Identification account, to be able to access Internet services, but have different accounts for access control and accountability services. Unauthenticated users may hack into a home wireless network while being in the neighborhood. They may use wireless hacking tools to crack wireless networks. From accountability/liability perspectives, and if those hackers committed a crime using victim network, forensic investigators may struggle to trace back the crime to the hacker.

Many business requirements support also the separation of AAA accounts. Ultimately, home users can offer Internet services to nearby users without causing security concerns or impacts.

Connecting to Unsecured Wireless Networks

You could connect to an unsecured network, and the data you send, including sensitive information such as passwords and account numbers, could potentially be intercepted. Many attackers can possibly create "free WiFi" networks to be used as honey-bots. They can provide users with free internet access while intercepting and spying on their sensitive data.

Wireless Networks and Platforms Security Issues

Wireless transmissions are not always encrypted. Information such as e-mails sent by a mobile device is usually not encrypted while in transit. In addition, many applications do not encrypt the data they transmit and receive over the network, making it easy for the data to be intercepted. For example, if an application is transmitting data over an unencrypted WiFi network using http (rather than secure http), the data can be easily intercepted. When a wireless transmission is not encrypted, data can be easily intercepted by eavesdroppers, who may gain unauthorized access to sensitive information (e.g. host computers) without the need to be host administrators, power users or even users in those local hosts. Their root level role implies having an administrator privilege in all network or system resources.

Role-based access control (RBAC) is a popular access control model for kernel security control enforcements. Many of the policy-based security systems adopt RBAC when writing, and enforcing security roles (NIST 2010 report). The report showed that RBAC adaptation continuously increases between the years 1992 to 2010. The NIST RBAC model is defined in terms of four model components: core, hierarchical, static separation of duty relations, and dynamic separation of duty relations.

Transport Layer Security (TLS)

Communication between mobile device and service provider servers must be encrypted to keep it safe from hacking and man in the middle attacks. The protocol, Transport Layer Security (TLS) or Secure Socket Layer (SSL) provides encryption for such communication in mobile phones. In iOS, when Application Transport Security (ATS) is enabled, it ensures that TLS encryption is activated for all mobile communications. ATS is enabled by default for applications that are built for iOS. Recently Apple made ATS as part of the application approval process.

TLS can be also used to create a Virtual Private Network (VPN). This integrates authenticates tasks with those of authorization.

Remote Wipe Feature

A user of an Apple mobile phone can have the ability to wipe their phone remotely using a mobile device management (MDM) server and iCloud or ActiveSync. All files will be then inaccessible and new user may need to create a new encryption with their new OS installation. System can be also automatically wiped after several unsuccessful PIN attempts.

Location-Based Services and Privacy Control

Any mobile with GPS capabilities can run "location-based services". Those types of services map or correlate a GPS-acquired location of the mobile to local businesses, users, services, etc. A large number of applications developed for mobiles provide services that are location-dependent. By default, those applications can be allowed to acquire information about the current location of the users or their mobile devices. However, users can select which applications to be allowed to have access to location information. For example, map or GIS-based applications (e.g. iOS maps, Google maps, etc.) greatly depend on the mobile location. They can provide limited services if they are not allowed to access location information. Careful users may want to balance between services and security or privacy to decide which applications to allow. Users may also choose to switch off those services in certain time or when they are in sensitive areas (Fig. 8.5).

Fig. 8.5 Enable/disable
location-based services

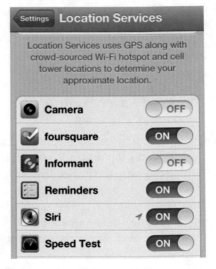

Skills Section

Software Security Issues in Smart Devices

- Download Lookout mobile security software on your mobile phone. Then create an account in Lookout website. Then show, in addition to malware scanning services, other services that Lookout provides (e.g. find my phone, contacts' backup, etc.). Make a report demo to show the overall process (i.e. installation and services' demo).
- Download a mobile application that can be used for mobile vulnerability testing (e.g. iScan online, Firewalla, SYOD, etc.). Then demo the report results from the application scan of your mobile.
- SYOD mobile application shows 9 security related settings that you should enforce to protect your device (passcode lock, auto-wipe data, device auto-lock, remote wipe, Safari settings, Siri Bypass, Wi-Fi, Disable Bluetooth, and Jailbreak). Based on the instructions in the application, show how you activate or enable those 9 features in your mobile phone (or whatever available in your mobile phone). Present a report with screen shots to show how to activate each security option.
- There are several websites (e.g. https://vulnerabilitytest.quixxi.com) that can test your mobile application for possible vulnerabilities (before making it public. Download or use one mobile application and test it in one of those websites for possible vulnerabilities. Show the result report in your task submitted document focusing on vulnerabilities if they exist in your tested application.

Smart Devices Operating Systems and Security Issues

- Based on your mobile device, download one of the spy mobile applications. Then demo how you can use it and access data online. Examples of top spy mobile applications include: for IPhone: mSpy, XnSpy, Mobistealth, FlexiSPY, Highster Mobile, Spyera, TheOneSpy, Auto-forward. For Android: for IPhone: mSpy, Mobistealth, FlexiSPY, Highster Mobile, Spyera, TheOneSpy, TeenSafe, etc.
- MetaSploit is a very popular vulnerability testing tool. There are some specific modules in MetaSploit to test mobile operating systems and applications. Submit a report that summarizes using MetaSploit to test your own mobile device.

Applications Section

Evaluating Access Controls in Smart Devices and Wireless Networks

Design a high level security control system to protect a mobile device. Describe 3–5 major policies that you propose in your policy system. Using (http://platform. screendy.com/), show basic UI screens on how to design the work flow in your proposed policy system (You may learn from the design of other policy systems that serve similar goals). Produce a report to your instructor to show the steps you accomplished in this task. As an alternative to an online IDE, you can use mobile IDEs such as xCode, Visual Studio, Eclipse or Android Studio.

Questions
- Make a comparison between iOS and Android in terms of security controls.
- Make a comparison between iOS and Android in terms of security vulnerabilities.
- Specify examples of how location-based services can impact security and privacy issues.
- Describe the meaning if "self-signed applications"
- How does the application code signing process help in detecting malwares?
- Describe a business or an information system. Then propose 2–4 examples of policies to control BYOD in this business or information system.
- Show with a demo example, how iPhone cloud lock works.
- Why it's better to use 6-digitd PIN code than 4-digits?
- If you lost the PIN code in iPhone can you reopen your iPhone? Can you recover data? Elaborate in both cases.
- Describe how mobile malwares could be different from regular desktops/laptops.
- Describe the different factors that malwares can be classified upon.
- In addition to the examples of malwares described in this chapter, research in the Internet for one more mobile malware and make a short summary on the malware, how it access victim machines, propagation, payload, etc.?
- Describe 3 main advices to avoid mobile malwares
- Do mobile anti-malwares have unique features from regular computers anti-malwares?
- Describe the main layers in iOS security architecture.
- Describe the role of embedded UID/GID in iOS security architecture.

- What is Jail-breaking? What's wrong with trying to jailbreak your phone?
- Why mobile operating systems keep pushing updates? Why it's important to keep your mobile OS up to date?
- Will you enable remote wipe? What are advantages and disadvantages of doing that?

Chapter 9
Software Code Security: Lesson Plans

Competency: Learn major aspects of software code security and vulnerability issues
Activities/Indicators
• Study reading material provided by instructor related to major aspects of software code security and vulnerability issues
• Complete successfully an assessment provided by instructor related to competency content.
• For mastering levels, more than 80% of assessment grades should be earned in no more than three trials.
• Assessment questions can be pulled from the end of the chapter questions or any relevant material.
Competency: Learn major aspects of software penetration testing
Activities/Indicators
• Study reading material provided by instructor related to software penetration testing.
• Complete successfully an assessment provided by instructor related to competency content.
• For mastering levels, more than 80% of assessment grades should be earned in no more than three trials.
• Assessment questions can be pulled from the end of the chapter questions or any relevant material.
Competency: Learn about secure software design principles and practices.
Activities/Indicators
• Study reading material provided by instructor related to secure software design principles and practices
• Complete successfully an assessment provided by instructor related to competency content.
• For mastering levels, more than 80% of assessment grades should be earned in no more than three trials.
• Assessment questions can be pulled from the end of the chapter questions or any relevant material.

(continued)

© Springer International Publishing AG 2018
I. Alsmadi et al., *Practical Information Security*,
https://doi.org/10.1007/978-3-319-72119-4_9

Competency: Learn principles of software secure construction and defensive programming, exceptions and error handling

Activities/Indicators
- Study reading material provided by instructor related software secure construction and defensive programming, exceptions and error handling
- Complete successfully an assessment provided by instructor related to competency content.
- For mastering levels, more than 80% of assessment grades should be earned in no more than three trials.
- Assessment questions can be pulled from the end of the chapter questions or any relevant material.

Competency: Learn on code analysis: Static and dynamic analysis

Activities/Indicators
- Study reading material provided by instructor related to static and dynamic analysis.
- Complete successfully an assessment provided by instructor related to competency content.
- For mastering levels, more than 80% of assessment grades should be earned in no more than three trials.
- Assessment questions can be pulled from the end of the chapter questions or any relevant material.

Competency: Learn on software malware analysis

Activities/Indicators
- Study reading material provided by instructor related to software malware analysis.
- Complete successfully an assessment provided by instructor related to competency content.
- For mastering levels, more than 80% of assessment grades should be earned in no more than three trials.
- Assessment questions can be pulled from the end of the chapter questions or any relevant material.

Competency: Learn how to conduct static software security code reviews

Activities/Indicators
- Use one or more open source or free tools that allow users to evaluate static software security code reviews
- For mastery, student is expected to try more than one operating system.
- For a mastering level, student should show what tool they selected, how they installed the tool and/or present a video to demonstrate the different commands they have used.

Competency: Learn how to test software for vulnerability issues

Activities/Indicators
- Use one or more open source or free tools that allow users to test software for vulnerability issues
- For mastery, student is expected to try more than one DBMS.
- For a mastering level, student should show what tool they selected, how they installed the tool and/or present a video to demonstrate the different commands they have used.

Competency: Learn how to use tools for code static and dynamic code analysis.

Activities/Indicators
- Use one or more open source or free tools that allow for code static and dynamic code analysis
- For mastery, student is expected to try more than one website or web server.
- For a mastering level, student should show what tool they selected, how they installed the tool and/or present a video to demonstrate the different commands they have used.

Competency:
- Learn how to conduct malware analysis in sandboxing environments.

(continued)

Activities/Indicators
- Use one or more open source or free tools that allow users to evaluate malware analysis in sandboxing environments
- For a mastering level, student is expected to show more than one RBAC example or application.
- For a mastering level, student should show what tool they selected, how they installed the tool and present a video to demonstrate the different commands they have used.

Competency: Learn how to evaluate software programs for security vulnerabilities.

Activities/indicators
- Conduct a study to evaluate software programs for security vulnerabilities.
- For a mastering level, student is expected to show more than one OBAC example or application.
- For a mastering level, student should show what tool they selected, how they installed the tool and present a video to demonstrate the different commands they have used.

Competency: Learn how to design a software security assessment system based on security requirements.

Activities/indicators
- Conduct a case to design a software security assessment system based on security requirements
- For a mastering level, student is expected to show more than one OBAC example or application.
- For a mastering level, student should show what tool they selected, how they installed the tool and present a video to demonstrate the different commands they have used.

Overview

In information systems, software programs are major components that exist to implement services/functions. They integrate hardware, network and environment to provide services for users that can vary based on the software goal.

The technology trends continuously shift toward the software and automation. This means that many services that users used to do manually are gradually moving to be programmable. Users gradually lose more share to software programs in the management and control of information systems.

Software programs are usually seen as the most vulnerable component from security perspectives in any information system. They are viewed as "easy-targets" that attackers usually focus on.

In this chapter, we will focus on security issues in software programs. We will cover software programs in their broad sense which may include programs in desktop, mobile and web platforms.

Knowledge Sections

Developing a secure software code and programs has been one of the major subjects in software design and quality. Details, methods and goals can vary from one programming language or environment to another, or from one platform (e.g. Desktop, Web, Mobile) to another or even from one application to another (based on the required levels of security, privacy, etc.).

In this section, we will visit in breadth software security aspects in all those variations of programming languages, environments or security levels.

Software Code Security and Vulnerability Issues

Vulnerabilities in a developed program can be traced back to different types of sources. Those include: The programming language in which the program is developed in, the program itself and how software developers design and construct the code, the platform or environment in which the software is deployed in and also the users who are going to use the software. In this section, our focus is only in the first part or the vulnerabilities that are related to the programming language or that the programming language can have an impact in.

Buffer and Stack Overflows

Applications are loaded to memory when they are under usage/execution. A buffer overflow exists when an application tries to write data beyond the buffer region. A buffer is a reserved memory space for a program variable. Buffer overflows can occur from programs' errors but they can also be crafted intentionally by hackers. As a result, an application can crash attackers may gain system access without proper authorization or may achieve a privilege elevation. Applications' memories are stored in either the application stack or the general heap storage.

Overflow attacks occur when attackers manipulate data input to a program. Such data manipulation causes problems to temporary data stored in memory. In overflow attacks, attacker tries to write data beyond the location that software designers intended for the input data. A memory stack is a memory location that is only accessible from the top. It is used to store local variables and function addresses. In a stack, a push operation inserts new data to the top of the stack and a pop operation removes data from the top of the stack.

Ultimately, writing secure software code (e.g. with only safe constructs) can ensure avoiding overflow attacks. For overflow attacks in particular, programmers should have code to check (not only assume) that data values entered by users will not go beyond the limit decided by the program.

Programming languages are different in their interaction/response to overflow attacks. Some programming languages enforce the use of only safe programming constructs.

Consequences of overflow attacks can be either to generally cause code, data corruption or program failures. Alternatively, in some cases, a successful overflow attack may make attackers succeed in creating program or system backdoors or execute their own malicious codes.

Memory Leak and Violation Issues

When users launch a program, the program will be loaded from the disk storage to the memory. Different programs can consume different memory sizes. Once users close those programs, they will be cleaned from memory to enable other applications to reclaim their memory and other allocated resources. This ensures better system performance and memory utilization. One of the main features in computer hardware that is continuously increasing is the memory size. As a consequence, users may not notice memory leak issues as previously unless if such leak is significant.

Different mechanisms exist in operating systems and programming languages to clean up memory and de-allocate memory resources that are not needed any more. A memory leak occurs as a failure to do that where mechanisms to de-allocate memory resources have some problems. In some cases, those different mechanisms may ignore memory cleaning as they assume that other mechanisms will do that.

Attackers, knowing that an application may have a memory leak, will try to send frequent requests to the application. As the application lacks a proper method to clean memory after each service call, this may ultimately cause a memory leak. As a result, application may deny further legitimate users services. While this problem is classical and has been in the security vulnerabilities for decades, recent security reports showed that memory leak problems still exist in software applications, operating systems and network devices (e.g. routers, switches, etc.).

The following are popular examples of memory schemes or mechanisms (i.e. for allocating and de-allocating memory):

- Heap memory allocation (programming constructs: HeapAlloc, malloc, new, etc.
- Operating system allocation (e.g. VirtualAlloc function).
- Kernel handles or memory (e.g. Kernel 32 APIs).
- User handles using User32 APIs.

Unlike C/C++ Java uses automatic garbage collection to automatically clean memory resources that are not used anymore.

Invalidated Inputs (SQL Injection and XSS)

Good software design principles require software developers to plan not only normal usage scenarios, but odd and exception scenarios. Users, intentionally or unintentionally may try to enter improper inputs to the software they are using. Desktop, mobile or web applications should be developed with the ability to handle properly responding to those incorrect or invalidated inputs. Programs should not only accept and process valid inputs, but they should also reject invalid inputs and do not process them further in the application. For example, SQL injection types of attacks depend on whether the evaluated website will accept and process invalid inputs.

Typically, websites include backend databases that store sensitive data that attackers are trying to access, expose, tamper or damage.

While many believe that current software security measures can counter SQL injection-based attacks, recent reports showed that such security problems still exist and can cause serious damages. Through the knowledge of how SQL statements are formed in the front end of web applications, attackers try to manipulate their inputs to trick backend databases. Based on the nature or the goal of the attack, attackers may try to expose the whole database, one or more tables, or particular certain fields or records in a table. They may also try also to view, change or destroy data in those databases.

An attack that tries to exploit a website for SQL injection typically performs the following three consecutive steps:

1. Scan the website for input forms or web elements that accept user inputs. Those are the initial candidate elements for SQL injection. Sometimes websites may hide those input elements and will not be visible to regular users. This is done to possible protect from SQL injection attacks but also as those input elements may have to be accessed by other systems or programs and not directly by users.
2. SQL injection attacks or testing tools then try an inventory of known (invalid inputs or categories of inputs, Table 9.1). A vulnerable website can possibly be vulnerable only to one or more types and hence it's important to evaluate all those categories of invalid inputs.
3. Once an input form is detected to be vulnerable for one category of invalid inputs, attackers or vulnerability testers can further exploit this type of attack and try to reach to the backend database.

Cross Site Scripting (XSS) attacks injects malicious scripts into victim web pages. Different types of software flaws may make web pages vulnerable to XSS. The basic

Table 9.1 A sample of results in pages that indicate possible SQL-injection vulnerable pages

mysql_num_rows()	VBScript Runtime
mysql_fetch_array()	ADODB.Field
Error Occurred While Processing Request	BOF or EOF
Server Error in '/' Application	ADODB.Command
Microsoft OLE DB Provider for [ODBC Drivers error]	JET Database
error in your SQL syntax	warning: mysql_fetch_row()
Invalid Querystring	Syntax error
OLE DB Provider for ODBC	include()
mysql_fetch_assoc()	GetArray()
mysql_fetch_object()	FetchRow()
mysql_numrows()	Input string was not in a correct format
warning: mysql_result()	error in your sql syntax
warning: mysql_connect()	unclosed quotation mark
microsoft jet database engine	Unclosed quotation mark before the character string
Microsoft VBScript	You have an error in your SQL syntax

common characteristic between those flaws is that the input from the user is executed or processed without proper validation, for possible malicious content, before that execution. The term scripting in XSS as the malicious content takes the form of a Java, or any other script while it may also contain other types of web content (e.g. HTML, Audio, Flash, etc.).

XSS is popular in attacks such as: Man in the Middle (MiM) and session hijacking where an attacker uses XSS to steal an open session (e.g. session cookies or information) for a user and claim their identity for their bank or an e-commerce website. Attackers can then steal private data related to user account, and use it for further identity theft related crimes. While typically XSS malicious code exists between <script></script > tags, there are many examples of other examples of XSS attacks (i.e. based on Java script, or any other scripting language, execution constructs).

XSS attacks can be fall into two categories:

- **Stored or persistent attacks**

In this case, malicious scripts are stored permanently in victim servers. Users need to scan their computers to find and deal with the exploit.

- **Reflected attacks**

In reflected XSS attacks malicious scripts are not permanent. They are sent to victim users through a different route than the website they are visiting (e.g. an email link, a message, etc.). The browser executes the code as its coming from a trusted user.

Race Conditions

This problem typically happens in distributed systems or multi-threaded programs. In race condition, a software process output is dependent on the timing or sequence of other external or uncontrollable events. The problems particularly occur when the process output created undesired or expected results as a consequence of irregularities of one or more of the process inputs. From a security perspective, this concurrency of inputs or events can be a source of vulnerability or exploit. One example of security problems that fall within race conditions is called (time of check to time of use, TOCTTOU or TOCTOU). This problem is a consequence of system/program change between the time when a condition is checked and the time when the results of that checking process is used. The assumption is that in this time difference, something relevant occur which compromise the integrity of the process.

Race conditions may cause a service to be locked temporary or permanently (e.g. dead- or live-locks.). As a result, Denial of Service (DoS) will occur and further, attackers can use this incident to commit attacks or create backdoors.

Designing and testing code to avoid race conditions can be hard and complex. Lab or experimental settings can be different from operation environments. Testers can possibly simulate some but not all those different possible occurrences of race conditions in that may occur in real time. For web applications, interactions or inputs can come from the users in the client side where they may intentionally or unintentionally provide incorrect inputs or be slow in their response. The Internet and the network are also important contributions to this issue as you can never guess how much a request sent through the network or the Internet will take time to reach destination. The third important input contributor to a web application is the interaction with the database (e.g. same website backend database, a bank, credit agency, etc.). Most e-business interactions require inputs from those three major contributors.

Ideally making the code "autonomic" can help in reducing occurrences of race conditions. This means that developers should make codes to modify and check data within one block. In this case, any change will impact the whole block including both locations. Practically, many real use cases do not accept those two pieces of code to be in the same block. Assume for example, a global variable (although global variables are not recommended in general) that is read by many pieces of code in different locations. Those different blocks of code may expect to read different values from this global variable as such variable can be modified outside their context.

Software static and dynamic code analysis tools (e.g. Valgrind, that will be discussed in another section) can be used to detect the possible occurrence of race conditions.

Software Vulnerability/Penetration Testing

Vulnerability/penetration testing methods, techniques and tools are used to test software mutiny against security attacks. While in many cases, those two terms refer to similar goals, vulnerability testing implies that main focus is to test whether target software, information system, website, etc. has internal vulnerabilities or weaknesses based on some predefined standards. On the other hand, regardless of whether such weaknesses exist or not, penetration testing methods try to different methods to attack and break into the target software. Scanning the software for possible vulnerabilities is typically the first stage of any penetration testing or even hacking schemes.

In vulnerability testing, tools may test websites, information systems. This will be described in details in another chapter in this book. In this part, we will focus on testing software source code for possible vulnerabilities.

Static and Dynamic Security Analysis

A software code can be tested statically or dynamically for possible security problems. Static code testing looks into the source code rather than its binary or executable version. Static code analysis can be conducted manually by human testers or

Table 9.2 Software code analysis confusion matrix

TN: Analysis tool successfully verified that a code element (e.g. a method or a class) has no security problem	FP: Analysis tool results showed a security problem in a code element. Once tested manually security problem does not exist
TP: Analysis tool successfully identified a security problem in a code element (e.g. a method or a class).	FN: Analysis tool showed that a code element has no security problems. Once tested manually users found security problems in the code element

tools. On the other hand, dynamic testing is accomplished through running the binary or executable version of the code. Finding a tool that can automatically search for and find vulnerabilities is much more convenient and efficient that manual assessment. The portion of software code automation is continuously growing with developing new tools for static and dynamic code analysis tools. However, most of large scale tools are commercial products. Some security problems such as those related to authentication, access control, encryption, etc. are still hard to find through tools.

Security problems that come from the software requirements or design are hard to detect by software tools. National security auditing organizations continuously publish security guidelines. Software vendors are expected to provide evidences that their developed software conforms with those guidelines.

The continuous evolution of information technology products (i.e. software, operating systems, programming languages and tools, etc.) present another form of challenge for software security auditing and tools. However, this is an evolving process and tools are getting better with time.

The accuracy of detection of security problems is another challenge with software security analysis tools. Table 9.2 below shows the 4 elements (i.e. TN, TP, FP and FN) of the confusion matrix and their indication in the scope of security code analysis tools.

Ideally, we are looking for a security analysis tool that can achieve the followings: TN and TP 100%, FP and FN 0%. Further, TP and FP complements each other (i.e. TP + FP = 100%) and TN and FN complements each other (i.e. TN + FN = 100%). Realistically, different tools can show different percentages of false detections (i.e. FP and FN). As a result, users may need to test with more than one tool to have more confidence in results. Upon certain alarms, they may also need to conduct focused manual testing or auditing.

Examples of free or open source code security analysis tools include: FindBugs, PMD, SonarQube, Brakeman, Codesake Dawn, FindSecBugs, Codacy, ABASH, Flawfinder, Vega, Google CodeSearchDiggity, PreFast, RIPS, VisualCodeGrepper, BOON, JLint, OWASP LAPSE, and Xanitizer.

Examples of commercial code security analysis tools include: HP Fortify and cAdvise, IBM AppScan, Coverity, Parasoft, Parasoft Test, JTest and doTTest, Veracode static analysis, BlueClosure BC Detect, ApexSec, Astree, BugScout, CxSAST, Julia, KlockWork, Kiuwan, PVS-Studio, Sentinel Source, Seeker, and Source Patrol.

NIST (USA National Institute of Standards and Technology: https://www.nist. gov/) keeps an online list of most current source code analysis tools with basic details on each tool:

(https://samate.nist.gov/index.php/Source_Code_Security_Analyzers.html). Wikipedia keeps also a record divided based on the tool programming language: https://en.wikipedia.org/wiki/List_of_tools_for_static_code_analysis

Not only for security testing, but static and dynamic software testing tools are used to test different software quality attributes (e.g. performance, usability, accessibility, reliability, etc.). Those different software quality attributes interact with each other. Sometimes they may contradict with each other where for example increasing requirements to optimize a quality attribute (e.g. security) may result in a trade of or decreasing quality requirements for performance.

While static code testing is usually conducted manually by users and hence is more expensive that dynamic testing, static testing can cover more flaws related not only to software code, but its design or requirements. As such, while it's more expensive, but it can be considered as more cost efficient in many situations. This is similar to testing in general and the difference between manually and automated testing. Additionally, dynamic security testing is usually categorized as a black-box testing, looking at the executable without looking into the code as static security testing. Recently a third type is defined: Interactive Application Security Testing (IAST) which can be considered as a hybrid approach between static and dynamic application security testing. An example of IAST includes software security testing tools from Acunetix.

Secure Software Design Principles and Practices

Security standards exist to guide software companies on how to comply and make sure that developed software applications are not vulnerable for the different types of security threats. Just like any other software quality aspect, the majority of security properties/issues are tested after software code is developed. Security problems related to requirements are design will go to production and will be expensive to discover and fix after code development. On the other hand, security testing for software requirements and design are complex, expensive and largely manual. In most cases, companies ignore and bypass this testing stage due to time and resource issues. They think that eventually in the testing stage any problem can be discovered and fixed.

Many organizations like US NIST continuously produce and update Security related standards and guidelines. Baldrige: https://www.nist.gov/baldrige/how-baldrige-works/about-baldrige, is a recent example. Baldrige performance excellence program is designed to enable organizations to make their self-assessments. In this section, we will describe general software security principles:

- Continuously or frequently conduct security or vulnerability assessments. In the current information technology environment, things change very quickly. Any software code is continuously evolving with new code as a response to bug fixes, adding new features, etc. The environment in which the software is operating is also continuously changing. Hence security testing should be a process with a frequent scheduling to run. The frequent of this testing can vary based on the important of the software or the sensitivity of the data and the environment. Such vulnerability testing can also run after work without impacting the working time or environment. If company has enough resources, it should try to mitigate all types of vulnerabilities. Several low level vulnerabilities can lead to a serious security vulnerability. Professional hackers know how to test for those vulnerabilities and find them. They know also how to expose systems based on whatever privileges they can get from the system.
- Based on the availability of resources, fix vulnerabilities according to their seriousness or risk level. The results of vulnerability testing should be continuously monitored. The goal is not to run those tests but to act based on the results. If time or resources are limited, at least make sure to fix or mitigate serious vulnerabilities. Making your company or information system a hard target for hackers can save you serious consequences and impacts.
- Layering and redundancy: Security architectures and mechanisms should always consider a layered architecture in which each layer can have its own security testing or guard mechanisms. Sensitive resources should, whenever possible be located in the inner layers. Redundancy in security controls can help ensure that systems are not compromised when one security control or layer fails to provide its expected protection.
- Plan for mitigation and backup. In information technology in general and security in particular, things may not always work as planned or expected. Nonetheless, systems should not reach an expected state where things are not accounted for. A security assessment process should many alternatives and "what-ifs".
- Privilege control: Many design principles exist to regular how to design security access control. The "least privilege" principle suggests that by default, users, applications, roles, etc. should be given least privileges. Unless required higher level privileges should not be granted to anyone of those listed earlier. Similarly, "least knowledge or least visibility" requires users are roles to be blocked from reaching resources that they don't have any interaction with. Separation of concerns in access control indicates the need to have different roles for different functions. Users that play different roles in the system should not be allowed to always use higher level privileges. This separation also requires that those who grant permissions to be separated from those who own or use those permissions.

Those and many other access control roles require the need to have a robust access control system that can allow making controls with fine-grained level of details on system resources.

- Trust and security are not built on assumptions but mechanisms. A resource is assumed secure only if there is at least one explicit security mechanism to protect it.

Similarly, assume in this current environment that "someone" exists who is interested to find and exploit your vulnerabilities. As such, it's better that you find and deal with those vulnerabilities.

- Test for insiders as much as you test for outsiders. Security reports showed that most security breaches started or got help, intentionally or unintentionally, from system insiders. Most security architectures explicitly include security controls for externals and largely ignore testing controls and auditing for internals. While detecting and mitigating against insiders' attacks that are committed intentionally by insiders can be very hard, systems should at least enforce policies to protect from unintentional insider attacks. For example, password security enforcements should be implemented to prevent users/employees from creating simple passwords that can be easily cracked by hackers.

Common Software Security Design Flaws

Security tools that check for vulnerabilities usually found vulnerabilities related to code implementation. Design flaws can be considered as roots or indirect cause to many of code vulnerabilities. For example, lack or improper methods of inputs validations are design flaws that cause many security vulnerabilities related to buffer overflow, SQL injection, etc. Similarly, access control or encryption design flaw or improper design can create different types of vulnerabilities and exploits. Literature includes several reports on top or most frequent or important software security related design flaw to avoid. Here is a summary list:

- **Validate user inputs**: All software applications require users' interactions to respond and include their inputs. User inputs can come in different forms from the very free text form (where users can type anything they want), to some restricted forms where users will pick from. Ideally, and whenever possible, limit the amount of data to input by users.

If possible, make all or most of those as select from alternatives where users will not have to type any free text. However, if functionalities require users' free input text, validate this input and don't assume users are going to follow generally accepted roles. Users, intentionally or unintentionally may try to violate those roles. Validate user input as early as possible. There are many possible designs. In some cases, especially in web applications users can be prevented, at the user interface level, from trying to include invalid inputs or characters. For example, if the form does not accept more than 5 characters, even if users try to type more than 5 characters, user interface will not allow them. As an alternative, inputs can be validated in the code rather than the user interface. Mainly, this input should not be further processed by code and certainly not be allowed to reach back-end databases, without being properly validated.

In some types of Man in the Middle attacks (MiM), attackers may try to attack and tamper data in transit. This means that proper input data validation should not be only employed to inputs directly taken from users, but to data inputs in general.

Some web applications may receive inputs from other websites (e.g. a bank, credit agency, government agency, etc. website). Input data should be validated from both trusted and un-trusted sources.

- **Use encryption properly or correctly**: Encryption should be used whenever possible. Encryption is used in many different places in software applications and information systems. It is used to hide information in rest (i.e. in files and databases) and is also used to hide information in transit. In some scenarios, encryption is required to hide all communication content between two parties. In some other scenarios, only some parts of this communication (e.g. users' credentials) should be encrypted when in transit. Use the right implementation of encryption as merely using encryption does not guarantee data, identity and privacy protection.
- **Use proper authentication and authorization levels**: Software developers usually use and employ security control modules and APIs. Those are integrated within the environment and can be used across different applications. Software developers do not need to reinvent the wheels and in most cases just properly implement whatever security standards exist. Security control mechanisms exist in operating systems, access control systems, network components, database management systems, websites, software applications, etc. As such, the application security is one layer in this layered architecture that should be designed in alignment with this overall architecture. In some cases, a software application may not need to consider security modules internally but properly integrate with security components that exist in the environment. In security layered architecture, authentication and identity management comes before access control or authorization. Authentication tries to answer a yes, no question whether user, application, request, etc. can be allowed or blocked access request or entry. Once allowed access control mechanism should make further decision of what permission levels this request can have.

  In relation to this part, code for data processing should be separated from that of security control processing. Data can be manipulated by users or their input and hence this data should not be mixed with control to ultimately commit attacks such as those related to privilege escalation.
- **Design for future changes**: Many of the design problems arise when a program design is in continuous change. Initial designers may did a good job based on the initial scope. However, there was not clear plan for extensions. Extensions/modifications for software applications are inevitable. They will come either as part of maintenance, bug fixes, etc. They may also come in response to new features or due to changes in the environment where the software must accommodate those changes. Cosmetic design changes that may come in response to those changes may ignore important design principles.
- **Take careful and extra considerations when dealing with sensitive data**: Attackers start from weak or vulnerable targets. Software designers should do their best to harden their applications. Sensitive data can be easily located in any application by software developers as well as attackers. Hence, designers should put extra considerations for areas that include sensitive data and make sure through frequent auditing and testing that those sensitive areas are not easy targets to attack.

Fig. 9.1 An example of
exception handling.

```
while(){
  ...
  try{
   case...
   default:
     throw new IllegalArgumentException("Invalid input...");
  }catch(IllegalArgumentException iae){
   //do stuff like print stack trace or exit
   System.exit(0);
  }
}
```

Software Secure Construction and Defensive Programming, Exceptions and Error Handling

Software code exception handlings are code components that are called in abnormal, exceptional cases. The main goal of design for code exceptions is to stop software applications from crashing at run time. Exception handling techniques can be used to block invalid inputs from users. Figure 9.1 below shows a simple example of code exception handling.

Exceptions block invalid user inputs early in the code before further code processing and before passing those inputs to internal system components (e.g. Databases). Exceptions also serve important goal of communicating effectively with users to show them what went wrong or how to properly communicate with the application with valid inputs. This is an important "usability" software quality attributes. Users should know why their requests are not accepted or processed. Exception feedback should help users fix the problems with their inputs. However, exception feedback should not be "revealing". In many security attacks, attackers intentionally try different types of inputs and try to use feedback exceptions to learn more about the system. Recent security research projects focused on this issue and why software exception handling feedbacks need to balance between usability and security issues. A balance feedback ensures that exception handling can serve both software quality and usability goals.

Software Malware Analysis

Malwares are software applications built with malicious intents. They are sent to user desktops, laptops, mobiles or websites through different techniques such as tricking users to receive or download those Malwares through some social engineering techniques. Anti-Malware software applications can work either per request or in real time to detect such Malwares, alert users and block them from accessing computing resources.

Malware analysis can be conducted automatically by anti-Malware systems or manually by security and forensic experts. Due to performance and speed issues, anti-Malware systems employ lightweight quick detection methods in comparison with those complex detection and analysis techniques employed through manual analysis.

Anti-malware Detection Techniques

Three major detection techniques are employed by anti-Malware systems to detect Malwares:

- Signature-based: Anti-Malwares used simple and reliable signature-based methods to identify good applications and files from malicious ones. A popular signature-based method is hashing. Hashing algorithms such as SHA1 and MD5 can be used to generate unique Hexadecimal values for system files and folders. Figure 9.2 shows a simple file hashing example with values from different hashing algorithms.

Different hashing algorithms can generate different hashing values. Hashing methods used not only in Anti-Malware system, rather many applications use hashing for integrity checking. A robust hashing algorithm should have the following criteria:

- For the same hashing algorithm, the same file without any changes should always produce the same hashing value.
- Any small change [even a space] in the file or folder should cause a change in the hashing value. There is no relation between hashing value and the amount of change made in the file.
- Reverse engineering for hashing values is impossible. This means that in a particular hashing algorithm, knowing the hash value cannot help in retrieving all or even part of the original file or folder.

Collision may rarely occur in hashing algorithms. Collision indicates that the hashing algorithm generates the same hashing value for different files or folders.

Anti-Malwares keep records of hashing values for the "good" files and applications. Those hashing can be extracted from public sources. For example, National Software Reference Library keeps records of known applications: https://www.nsrl.nist.gov/. Malware scanners can make real time queries to those hashing databases. Alternatively, they can keep local records for faster queries.

Malwares try to avoid signature-based detection using different techniques. In some polymorphic Malwares, the Malwares can have different forms or behaviors to make it hard to confirm their identities. New Malwares can also be designed from old known Malwares with slight changes. Signature-based Malware detection will see those Malwares as different from earlier versions.

- Role-based: Hashing-based Malware detection can work well in known territories. However, for new Malwares, such Malwares will not be registered as known malicious

Results	
Original text	*(binary only)*
Original bytes	255044462d312e340a25d3ebe9e10a312030206f626a0a3c3c... (length=131655)
Adler32	d85d3d84
CRC32	121fb30f
Haval	a4445c97579f9069ba27a5672b504421
MD2	3ac77941484a0103f8b3b859b8c2a783
MD4	ed44d3b944827ef0bbbc28add906fca0
MD5	85125c2d3e499c7ab8af8c7e61102435
RipeMD128	d7ab86a75e84fbb948c2a6929c59fab7
RipeMD160	4acea0bdbb3ffae97c0de24510bac478b60eef7e
SHA-1	797c34e9a1752ee957c06762e8b42636a6b63850
SHA-256	0621e7dd8ad7a15fdcc842e157767fc0a14d8a9c0cebafc3c66d1d88c5f3af46

Fig. 9.2 An online tool for hashing.

files or applications. Role-based detection methods are based on the identifications of different behaviors that indicate that the subject file or application is a Malware. For example, an application that tries to access or change certain sensitive areas or information in operating system kernels can be suspected as a Malware. Apparently, this may cause a significant number of false positives, where normal applications are trying to trigger such requests. It may also cause cases of false negatives where some Malwares may not be detected as they are either trying to make hidden moves or their triggers are not recorded by any defined role.

- Behavior analysis: Malwares can be very complex. Detecting such complex Malwares may take more than a second or a part of a second in real time. In behavior analysis, certain sequence of activities may trigger a suspicious behavior. In some cases, this may require full packet states' inspection (Also called deep packets inspection) in order to be able to judge whether this can be a Malware or not. Certain library (e.g. APIs) or system calls can be also defined as Malware-like behaviors. Anomaly detection methods define ranges of acceptable behaviors. Any "deviation" from such behavior can be classified as "anomaly" and hence by a possible Malware.

  Malwares can be analyzed in isolated sandboxing environment. Here are some techniques that some Malwares may employ to complicate detection:

  – Encryption and compression: The detection of a Malware that is compressed, encrypted or both in real time is impossible. Special decompression and

encryption methods are required. Those may take time and may not always be successful or easy to accomplish.
- Obfuscation: Malwares need to be analyzed and reverse engineered for analysis. Some Malwares use obfuscation to prevent reverse engineering activities.
- Multi-partite or polymorphism: Malwares, to avoid detection may also use different methods to intrude or access victim machines. This may complicate detection methods.

Manual Malware Analysis

In forensic investigation for thorough Malware analysis , the automatic or real time detection can be only the initial stage. Other than detecting the Malware, investigators are interested to know more details about this Malware to prevent its further impact. As part of Malware analysis, investigators are interested to analyze the Malware to understand its access or intrusion methods. Each Malware has its way of finding vulnerability in information system and using it to access victim machine. Those weaknesses may not be all the time related to the software or hardware components of the information systems. They can be also in a form of social engineering technique where a victim user is tricked to open a malicious link or file.

In addition to how this Malware intrudes victim machines, forensic investigators are interested to know how such Malware spreads from one machine or file to another. They are also investigating the nature of the payload for the Malware.

Malwares can be complex to analyze. They may employ techniques to block or resist Malware analysis. For example, some Malwares may show polymorphic behaviors and act differently in different environments or different times. For example, some Malwares can be very active if they are in Windows hosts and dormant if they are in other operating systems. Malwares can also resist analysis using compression, obfuscation and encryption techniques.

Skills Section

Software Security Testing

The following two links list examples of many software security testing tools:

- https://samate.nist.gov/index.php/Source_Code_Security_Analyzers.html
- https://en.wikipedia.org/wiki/List_of_tools_for_static_code_analysis
  Select only one tool based on your preference. Make sure you can download and install the tool properly. Then demo using the tool to test one software application that can be properly tested using the tool and show the output report.

Static Software Security Code Reviews

Pick one static and one dynamic security code analyzer tools from those described in (Software vulnerability/penetration testing). Install the tool and demo installing and using it for at least one source code of your choice.

Student is expected to:

- Pick one of the tools from the above list (or any similar tool), based on their own programming language and environment preferences.
- Pick experimental software to test. This can be from their working environment or any other software that they like to test.
- Extract report from the tool and summarize result and submit it to the course lab work or present it in class.

Malware Analysis

You can use public hash datasets to search for possible Malwares in your system. National Software Reference Library keeps records of known applications: https://www.nsrl.nist.gov/. Download a hashing dataset from this website. Then install NSRL server in your system (https://github.com/rjhansen/nsrlsvr). Once server is installed and working properly, you can test it by hashing selected files and folders in your system (e.g. using md5deep hashing tool). Finally, you can compare hashes of your files with those extracted from NSRL using your installed NSRL server.

Use the following links to help you complete scanning your system for possible Malwares: 1. http://blog.jameswebb.me/2013/05/setting-up-forensic-hash-server-using.html, 2. http://sysforensics.org/2013/12/build-your-own-nsrl-server/. Show the final scanning report. For those who may struggle building their own local NSRL server, you can make queries to this publicly made server described in the link http://jessekornblum.livejournal.com/278435.html. The server exists in the address: nsrl.kyr.us. Hence for example you can make commands such as: md5deep -r * | nsrllookup -s nsrl.kyr.us rather than having your own local NSRL server.

Applications Section

Evaluating Software Programs for Security Vulnerabilities

Pick software, information system or a website either from your work or from your choice. You are supposed to submit a detail report on the security stance of this software, information system or website. Based on the nature of the system, you are supposed to select one or more proper vulnerability scanning tools. Summarize the results of those vulnerability scanning tools. You are also expected to conduct manual security assessment. Manual assessment can be made either based on initial tools' assessment for further clarifications. Manual assessment can also focus on areas that are very hard to check or detect by tools.

Your final report should also include recommendations on how to improve security in evaluating system. It should also include specific action tasks on how to fix discovered vulnerabilities and to make further security assessments.

Software Security Assessment Based on Predefined Security Requirements

Many standards exist to evaluate companies or systems from security perspectives. Two of those guidelines are:

- Federal Financial Institutions Examination Council (FFIEC) cyber security assessment tool: https://www.ffiec.gov/cyberassessmenttool.htm
- Baldrige Cybersecurity Excellence Builder: https://www.nist.gov/baldrige/products-services/baldrige-cybersecurity-initiative

Select one of the two and for each criterion, select your assessment based on your company or organization. Include a new column to justify your selection. If you don't have a company or you dont want to select your own company, you may evaluate the University as a company (or any company you want to pick and evaluate).

Questions
- Describe the difference between buffers and stack overflow attacks.
- Describe one mechanism to protect against overflow attacks.
- Describe one example from a programming language in which language constructs exist to protect against overflow attacks.
- What is the meaning of memory leak and why it happens?
- What is the security implication of memory leaks?
- Describe one example from a programming language in which language mechanisms exist to protect against memory leaks.
- How can Java garbage collectors help against memory leaks?
- Describe one example of an SQL injection.
- Select an SQL injection vulnerability from Table 9.1. Then show how you can test if such vulnerability exists and how to exploit such vulnerability.
- Describe one example of an XSS attack.
- Describe one tool that can be used to test against SQL injection.
- Described one tool that can be used to test against XSS attacks.
- What is the difference between XSS stored and reflected attacks?
- Describe "race condition" problem and its security impacts or implications.
- Use a tool to demo one instance of a race condition problem in a software you are testing.

- Make a table of comparison for the strengths and weaknesses of conducting static and dynamic security analysis methods.
- Describe with one example for each the four detection results in Table 9.2.
- How is IAST different from SAST and DAST?
- In addition to security design principles mentioned in this chapter, make your own selection of one security code/design principle that you think is important but not mentioned in this chapter. Justify your selection.
- Make a table of comparison between the different Malware detection methods describing weaknesses and strengths for each method.
- Validating user inputs is mentioned as an important code/design principle. Describe two techniques that can be used to accomplish such validation.
- How could code exceptions be useful for security?
- How could improper exception implementations hurt software security?

Chapter 10
Disk and Computer Forensics: Lesson Plans

Competency: Learn major aspects of disk forensics in FAT Systems.

Activities/Indicators
- Study reading material provided by instructor related to major aspects of disk forensics in FAT Systems.
- Complete successfully an assessment provided by instructor related to competency content.
- For mastering levels, more than 80% of assessment grades should be earned in no more than three trials.
- Assessment questions can be pulled from the end of the chapter questions or any relevant material.

Competency: Learn major aspects of disk forensics in NTFS Systems.

Activities/Indicators
- Study reading material provided by instructor related to major aspects of disk forensics in NTFS Systems.
- Complete successfully an assessment provided by instructor related to competency content.
- For mastering levels, more than 80% of assessment grades should be earned in no more than three trials.
- Assessment questions can be pulled from the end of the chapter questions or any relevant material.

Competency: Learn major aspects of disk forensics in ext. Systems.

Activities/Indicators
- Study reading material provided by instructor related to major aspects of disk forensics in ext. Systems.
- Complete successfully an assessment provided by instructor related to competency content.
- For mastering levels, more than 80% of assessment grades should be earned in no more than three trials.
- Assessment questions can be pulled from the end of the chapter questions or any relevant material.

Competency: Learn major aspects of disk forensics in HFS Systems.

Activities/Indicators
- Study reading material provided by instructor related to major aspects of disk forensics in HFS Systems.
- Complete successfully an assessment provided by instructor related to competency content.
- For mastering levels, more than 80% of assessment grades should be earned in no more than three trials. Assessment questions can be pulled from the end of the chapter questions or any relevant material.

(continued)

© Springer International Publishing AG 2018
I. Alsmadi et al., *Practical Information Security*,
https://doi.org/10.1007/978-3-319-72119-4_10

Competency: Learn major aspects of memory forensics.

Activities/Indicators
- Study reading material provided by instructor related to major aspects of memory forensics.
- Complete successfully an assessment provided by instructor related to competency content.
- For mastering levels, more than 80% of assessment grades should be earned in no more than three trials.
- Assessment questions can be pulled from the end of the chapter questions or any relevant material.

Competency: Learn major aspects of operating system forensics in windows environments.

Activities/Indicators
- Study reading material provided by instructor related to major aspects of operating system forensics in Windows environments.
- Complete successfully an assessment provided by instructor related to competency content.
- For mastering levels, more than 80% of assessment grades should be earned in no more than three trials.
- Assessment questions can be pulled from the end of the chapter questions or any relevant material.

Competency: Learn major aspects of operating system forensics in Linux environments.

Activities/Indicators
- Study reading material provided by instructor related to major aspects of operating system forensics in Linux environments.
- Complete successfully an assessment provided by instructor related to competency content.
- For mastering levels, more than 80% of assessment grades should be earned in no more than three trials.
- Assessment questions can be pulled from the end of the chapter questions or any relevant material.

Competency: Learn major aspects of operating system forensics in Apple MAC environments.

Activities/Indicators
- Study reading material provided by instructor related to major aspects of operating system forensics in Apple MAC environments.
- Complete successfully an assessment provided by instructor related to competency content.
- For mastering levels, more than 80% of assessment grades should be earned in no more than three trials.
- Assessment questions can be pulled from the end of the chapter questions or any relevant material.

Competency: Learn an example on how to conduct disk forensics in FAT Systems.

Activities/Indicators
- Use one or more open source or free tools that allow users to conduct disk forensics in FAT Systems.
- For mastery, student is expected to try more than one FAT system.
- For a mastering level, student should show what tool they selected, how they installed the tool and/or present a video to demonstrate the different commands they have used.

Competency: Learn an example on how to conduct disk forensics in NTFS Systems.

Activities/Indicators
- Use one or more open source or free tools that allow users to conduct disk forensics in NTFS Systems.
- For mastery, student is expected to try more than one NTFS system.
- For a mastering level, student should show what tool they selected, how they installed the tool and/or present a video to demonstrate the different commands they have used.

(continued)

Competency: Learn an example on how to conduct disk forensics in ext. Systems.

Activities/Indicators
- Use one or more open source or free tools that allow users to conduct disk forensics in ext. Systems.
- For mastery, student is expected to try more than one ext. system.
- For a mastering level, student should show what tool they selected, how they installed the tool and/or present a video to demonstrate the different commands they have used.

Competency: Learn an example on how to conduct disk forensics in HFS Systems.

Activities/Indicators
- Use one or more open source or free tools that allow users to conduct disk forensics in HFS Systems.
- For mastery, student is expected to try more than one HFS system.
- For a mastering level, student should show what tool they selected, how they installed the tool and/or present a video to demonstrate the different commands they have used.

Competency: Learn an example on how to conduct memory forensics.

Activities/Indicators
- Use one or more open source or free tools that allow users to conduct memory forensics.
- For mastery, student is expected to try more than one operating system.
- For a mastering level, student should show what tool they selected, how they installed the tool and/or present a video to demonstrate the different commands they have used.

Competency: Learn an example on how to conduct forensics in Windows Operating Systems.

Activities/Indicators
- Use one or more open source or free tools that allow users to conduct forensics in windows Operating Systems.
- For mastery, student is expected to try more than one Windows operating system.
- For a mastering level, student should show what tool they selected, how they installed the tool and/or present a video to demonstrate the different commands they have used.

Competency: Learn an example on how to conduct forensics in Linux Operating Systems.

Activities/Indicators
- Use one or more open source or free tools that allow users to conduct forensics in Linux Operating Systems.
- For mastery, student is expected to try more than one Linux operating system.
- For a mastering level, student should show what tool they selected, how they installed the tool and/or present a video to demonstrate the different commands they have used.

Competency: Learn an example on how to conduct forensics in MAC Operating Systems.

Activities/Indicators
- Use one or more open source or free tools that allow users to conduct forensics in MAC Operating Systems.
- For mastery, student is expected to try more than one MAC operating system.
- For a mastering level, student should show what tool they selected, how they installed the tool and/or present a video to demonstrate the different commands they have used.

Competency: Learn how to evaluate different file systems from disk forensics' perspectives.

Activities/Indicators
- Conduct a study to evaluate different file systems from disk forensics' perspectives.
- For a mastering level, student is expected to show more than two file systems.
- For a mastering level, student should show what tool they selected, how they installed the tool and present a video to demonstrate the different commands they have used.

Competency: Learn how to evaluate different Windows operating systems from disk forensics' perspectives.

(continued)

Activities/Indicators
- Conduct a study to evaluate different Windows operating systems from disk forensics' perspectives.
- For a mastering level, student is expected to show more than two Windows operating systems.
- For a mastering level, student should show what tool they selected, how they installed the tool and present a video to demonstrate the different commands they have used.

Competency: Learn how to evaluate different Linux operating systems from disk forensics' perspectives.

Activities/Indicators
- Conduct a study to evaluate different Linux operating systems from disk forensics' perspectives.
- For a mastering level, student is expected to show more than two Linux operating systems.
- For a mastering level, student should show what tool they selected, how they installed the tool and present a video to demonstrate the different commands they have used.

Overview

In digital investigations, many software and hardware components can be searched for possible relevant evidences. Digital forensic investigators should not only have knowledge on the subject case, but also on technical skills related to how to search for and acquire relevant information. Skills in disk and computer forensics continuously evolve with the evolution of computer hardware, software, operating systems and environments.

Books that include this subject show a large variation in the content of such subject. For example, in the Windows environment, early books cover: Disk Operating Systems (DOS), Windows 95, 98, XP, File Allocation Table (FAT) 16, and 32. More recent books cover recent Windows operating systems (e.g. Windows 7, 8 and 10), FAT32 and NTFS. Books also cover other operating systems and environments (e.g. Linux Ubuntu, Debian, Apple MAC) and their file systems.

Knowledge Sections

Searching for possible evidences related to an incident in a disk, a file or operating system can be a very time consuming and tedious process. This can be as a result of 3 factors:

- The number of possible files and applications to search within is typically very large. Initially, a forensic investigator is supposed to search within all files, folders and applications and not ignore any part. This can take a significant amount of time and resources. Further, this volume of data is continuously growing where a typical current operating system can have thousands or millions of files (Fig. 10.1).
- Forensic investigators are expected to study first the subject case so that they can search for what is relevant to the case. However, making such connection may

Fig. 10.1 An example of files/folders count in an operating system

require looking at every single detail without ignoring any piece of information. In some cases, significant evidences may possibly exist in places where many analysts will ignore.

- Disk and operating system investigation tools exist in the commercial, free and open source domains. Those tools also continuously change to accommodate disk and operating system changes. The possible value any tool can provide can vary from once incident to another. Digital investigators may need to try a large number of tools in every case. They may see different and sometimes conflicting types of information.

Disk Forensic Activities

Image Acquisition

Digital evidences are typically stored in different types of disks (e.g. desktop or laptop disk drives, USB drives, mobile drives, etc.). The first step in forensic cases that include one or more of those disks is to be able to acquire data from those disks and store them on a secondary storage. In most cases, investigators must not conduct their analysis on live disks. The main reason is to preserve the evidence integrity and verify that none of the information that exists in the disk evidence is added/edited/deleted by the forensic investigators themselves.

(dd) is one of the simplest Linux based tools to perform image acquisition:
dd if=/dev/sdc of = image.dd.................................. dd command for image
    acquisition.
/dev/sdc in this case represents the input disk to image copy or acquire. The output
    file (image.dd) represents a "raw" disk data format.
dcfldd is an advanced version of "dd" for forensic imaging. You can include more
    details on the imaging process such as adding different hashing algorithms and
    how to split the input image to several output partitions.
dcfldd if=/dev/sdc hash = md5,sha256 hashwindow = 10G md5log = md5.txt
    sha256log = sha256.txt \ split = 10G splitformat = aa of = image1.dd ....
    dcfldd command for image acquisition.

Raw image file formats have the advantage of being open source, simple and can be analyzed and used by many tools. However, little metadata can be extracted about the image using raw formats.

There are two major alternative formats: Commercial or proprietary formats and independent formats.

Commercial disk analysis tools have their own proprietary formats. For example, EnCase (.E01,. E02, etc), ILook (compressed (IDIF), non-compressed (IRBF), and encrypted (IEIF)), etc. Unlike raw image formats, proprietary formats extract metadata information, but can be only used within their specific tools.

As a compromise between the two previous options, Independent file formats (such as AFF, AFD and AFM) can be used across different tools while they also collect disk and imaging process metadata.

Investigators should be aware of those different data acquisition formats and tools. They need to take the proper of which acquisition format and tool to use for the current case they are investigating.

Hashing is used to verify the integrity of the disk and collecting evidences. There are many tools and online services to provide hashing. Hashing can be created for single files, folders or for an entire disk. Figure 11.2 show an example of an online website or tool that can be also used to create hashing hexadecimal codes (Fig. 10.2).

When a case is presented in course, forensic investigators should show that the current hashing values for the evidence disk match the same values when the disk evidence was acquired. Initial hashing values are stored with the evidence as part of its acquisition process. The same data should always create the same hashing value for the same algorithm (Fig. 10.3). However, any data change, even if it's as small as a single space, this should cause the generation of a new different hashing value.

Different hashing algorithms generate different hash values for the same data. Unlike encryption, there is no reversible link between output hashing value and input data. In other words, knowing a hash value can impossibly help in knowing the input data.

Results	
Original text	*(binary only)*
Original bytes	ffd8ffe2021c4943435f50524f46494c450001010000020c6c... (length=94841)
Adler32	7fb3b6a1
CRC32	9208beae
Haval	cd28f16055cbad8560f06ebb97edf0d4
MD2	98d7dad1c73a0189240683ffa3a8593f
MD4	c2e0c4d449d989ce88c33d056bc8c6ca
MD5	08ee31606c4cedc716da0b10329a
RipeMD128	bac154b92a027925ff93fc3b3ba51d71
RipeMD160	b363eed5bd2f6b079d7177c4ac36db05cb39f1ed
SHA-1	a42c8ba3acf44a35dcb206ee5be88ecf8c946339
SHA-256	d36f2f1c227cab5ee7d03a1ea33faf5f1c0b480315429020bd60012baa1accaa
SHA-384	272964d98b9bb39aea248d0a6167300e16baae6c49403a8a8fd2797751e06587140c3ba7c5e00eac0ac1a024816e08fc
SHA-512	2e89f38aa25cb1f1200655e6d08771d12c415556bb63f9d58b12b2aa566a1ccae790cd44ac56a5a9c5aa80fde686aec237ba0ef2db00bdb4636050d5562cf9d3

Fig. 10.2 An example of hashing with different algorithms

Fig. 10.3 The hash
verification (SANS 2011)

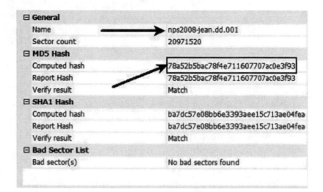

Data Recovery

Data recovery can be used as part of an investigation process or completely
independent. Data recovery can be a task allocated to personnel from information
technology or security department. A computer, server, website or database failure
can cause a corruption in data. Data recovery is the process to try to retrieve data to
an earlier state before such failure. In one classification, data recovery process can
be divided into forward and backward recovery:

- Forward recovery: The process of starting from the current, corrupted state and
  move forward to reach a state where corrupted data is fixed/treated. For example,
  if this is a Windows operating system that stopped from rebooting successfully,
  a forward recovery process will try to use tools to extract the data from the
  corrupted disk or fix the error to allow the operating system to boot again.

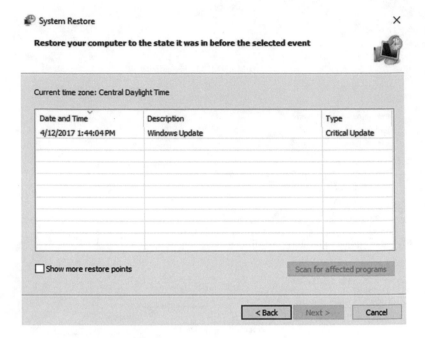

Fig. 10.4 An example of Windows restoration process

- Backward recovery: In some cases, where errors are not fixable or time-consuming to fix, users may settle with restoring back to an earlier stage where the system was working properly. Systems should be prepared to go back to those stable restoration points. Some recent data can be possibly lost in this backward recovery process. This is considered as a fault-tolerance method. Most operating systems currently provide such options. Users should enable such feature for going back to certain restoration points, Fig. 10.4.

Forensic Analysis

Data from disks are imaged in forensic processes so that they can be analyzed for possible evidences. The analysis activities can vary based on the nature of the case and the disks or images. Here are some generic activities that occur in most disk forensic cases:

- Hashing activities: In addition to creating and comparing hashes for disk files and folders, forensic tools can flag known files (e.g. system files) to be ignored from further search and analysis.
- Keyword search: This is one of the major tasks in this stage. In addition to generic search within the disk volume, files and folders, most disk forensic tools provide structured or predefined searched for keywords, Fig. 10.5.

Fig. 10.5 An example of keyword search

- Timeline analysis: As in most cases, forensic investigators are interested to focus their analysis in a window of time. Forensic tools can then help them aggregate all related files and activities within that border window.
- Artifacts: Forensic tools can provide artifacts such as emails, IP addresses, web links, software tools, etc. Such information can be extracted from all files and folders in the disk.

Disk Forensics in FAT Systems

Windows File Allocation System (FAT) originally started in 1977 to be used in floppy disks. Eventually it became the defacto file system for disks in DOS and Windows operating systems. It has the 3 variants FAT12, FAT16 and FAT32. The number after (FAT) indicates the number of bits used for cluster addressing. FAT allocates logical disk spaces called "clusters" to files based on their sizes. Files are known in FAT by their names and the location of the clusters they are reserving. Each FAT system has different cluster size in comparison with the physical size elements (i.e. sectors). FAT12 was proper for floppy disk or small size disk storages. Figure 10.6 shows different options for FAT12 disk sizes, sectors per cluster and cluster sizes.

Table 10.1 summarizes differences between FAT12, 16 and NTFS. Cluster size has a significant impact on the disk utilization and performance. When cluster sizes are large, disk will have more wasted space and less disk utilization. The size of a

Fig. 10.6 FAT12 disk
sizes

```
Drive size  Secs/cluster   Cluster size
   360 KB        2            1 KiB
   720 KB        2            1 KiB
   1.2 MB        1          512 bytes
   1.44 MB       1          512 bytes
   2.88 MB       2            1 KiB
```

Table 10.1 A summary of comparison between FAT16, FAT32 and NTFS

FAT16		FAT32		NTFS	
Volume size MB	Default cluster size KB	Volume size GB	Default cluster size KB	Volume size MB	Default cluster size KB
0–32	0.512	0.032 - 8	4	512 or less	0.512
32–64	1	8–16	8	512–1024	1
64–128	2	16–32	16	1024–2048	2
128–256	4	32	32	Greater than 2048	4
256–512	8				
512–1024	16				
1024–2048	32				
2048–4096	64				

cluster can vary from one sector or 512 bytes to 128 sectors or 65 K bytes. Maximum volume size for FAT16 is 4GB. This means that if you have a USB drive with a size more than 4 GB, you can't format it as FAT16.

The majority of file systems in current Windows machines will be either FAT32 or NTFS. However, forensic investigators may have to deal with some old disk storage formatted according to FAT 12 or FAT16.

Any FAT file system will consist of following 3 physical sections:

- A reserved area for file system information– This area starts from sector 0 or the first disk sector. This size of this area is given in the boot sector. The boot sector is the first sector in the reserved area.
- FAT area – For primary and backup FAT structures. One difference between FAT12/16 and FAT32 is that the root directory is fixed between the FAT and DATA areas in FAT12/16 while it can be anywhere in the data area in FAT32.
- Data area – Those are the clusters used for storing file and directory contents

File System Monitoring Tool (fsstat) can be used to make queries on file systems and file system layout. Information coming from this simple tool can also show file system type in investigated image or drive (Fig. 10.7).

Other simple open source tools such as mmls can be also used to learn about the file system type (Fig. 10.8).

Hexadecimal editor tools can be also used to detect the file system type. FAT file system has the values 0x55 and 0xAA in byte offsets 510 and 511 of the first sector.

Fig. 10.7 fsstat tool
example

```
# fsstat –f fat fat-4.dd
FILE SYSTEM INFORMATION
-----------------------------------------------
File system type: FAT
OEM Name: MSDOS5.0
Volume ID: 0x4c194603
Volume Label (Boot Sector): NO NAME
Volume Label (Root Directory): FAT DISK
File System Type Label: FAT32

Backup Boot Sector Location: 6
FS Info Sector Location: 1
Next Free Sector (FS Info): 1778
Free Sector Count (FS Info): 203836 ...

File System Layout (in sectors)
Total Range: 0 – 205631
*  Reserved: 0 - 37
** Boot Sector: 0
** FS Info Sector: 1
* FAT 0: 38 - 834
* FAT 1: 835 - 1631
* Data Area: 1632 - 205631
*** Root Directory: 1632 - 1635

CONTENT DATA INFORMATION
-----------------------------------------------
```

```
 mmls -t dos -vbr vm_forensics
tsk_img_open: Type: 0    NumImg: 1  Img1: vm_forensics
dos_load_prim: Table Sector: 0
tsk_img_read: Loading data into cache 3 (0)
raw_read: byte offset: 0 len: 65536
load_pri:0:0    Start: 63    Size: 16482627 Type: 7
load_pri:0:1    Start: 0    Size: 0  Type: 0
load_pri:0:2    Start: 0    Size: 0  Type: 0
load_pri:0:3    Start: 0    Size: 0  Type: 0
DOS Partition Table
Offset Sector: 0
Units are in 512-byte sectors

     Slot    Start        End            Length        Size    Description
00:  Meta    0000000000   0000000000     0000000001    0512B   Primary Table (#0)
01:  -----   0000000000   0000000062     0000000063    0031K   Unallocated
02:  00:00   0000000063   0016482689     0016482627    0007G   NTFS (0x07)
```

Fig. 10.8 mmls tool example

Forensic investigators may have to deal with different challenges with data that can't be seen or extracted using normal searching mechanisms. Here are examples of those challenges:

- Deleted data: Suspects may try to delete certain files that they feel can be used as evidences against them. When files are deleted, their records in the FAT table are deleted. However, the actual file information is not erased unless if a new content is added to the same ex-file location. Once files are deleted and their FAT addresses are claimed as "empty addresses" new files can then use them.

    The classical BTK killer case is an example of retrieving deleted files to be used as evidences. The suspect sent a floppy disk to officials in which they were able to retrieve his information from some deleted files.

- Hidden data: A hidden data or area in a disk is that data/area that is not seen by the file system. While most attempts in forensic cases can be related to either erasing or encrypting relevant data, however, it is possible that many occurrences of data hiding were not discovered. Here are difference examples of possible hidden areas in a disk:

    - Unused sectors in the reserved area
    - Slack spaces: There are different types of slack spaces that can be used to hide data including: File, RAM, drive, etc. File systems used fixed size containers called clusters to store files. The rest of the cluster at the end of the file is a file slack as this space cannot be used or claimed by other files. A file slack and the empty data between the last bit of file data to the end of the last cluster used by the file. Each file in the file system can have this "left over" and the total disk space can be the total slack spaces from all files. The RAM slack happens in memory as data is written in memory in sectors (blocks of 512 bytes). As such, last block in a retrieved file to the memory will be filled with random data to complete the last sector.

    Slack spaces can be used, from a forensic perspective in two aspects:

    - A professional hacker or suspect can possibly craft a malicious file, or application to be hidden in some or all different partition or disk slacks. While this may seem to be complicated, however, it is not impossible. On the other hand, such acts will be very hard forensic investigators to detect.
    - Slack spaces may keep data from earlier files. As a result, hackers may use tools to scan slack spaces looking for valuable information to steal. Users may assume that such data is deleted.
    - Between the end of the file system and the end of the volume
    - Data hiding using stenography and watermarking: Unlike encryption which disables readers from understanding data, while they can still read/see it, stenography and watermarking techniques are based on hiding some parts of data within other parts. Users with normal tools can see files or applications that they are supposed to see. On the other hand, other files or data portions are hidden within the visible data and can only be seen using certain methods and tools.

Table 10.2 Examples of file extensions and their start and end of filer markers

Extension	Start marker	End marker
jpeg	FFD8	FFD9
gif	47,494,638	003B
png	89504E470D0A1A0A	49454E44
html	3C48544D4C3E	3C2F68746D6C3E
pdf	25,504,446	2525454F46

- Every partition contains a boot sector. If the partition is not bootable, the boot sector in the partition is available to hide data.

• Corrupted data: In disk investigation, investigators may have to deal with corrupted data, whether intentionally or unintentionally. In addition to text analysis tools, hexadecimal editors can be used to deal with data corruptions. For example, file systems and files generated by the different applications have different hexadecimal tags that they can be checked. Table 10.2 shows examples of file types/extensions and their start and end of file markers. Those can be used to deal with corrupted files. They can be also used to deal with stenography and data hiding.

Unlike old hard disks, modern hard disks can handle bad sectors themselves by remapping bad sectors to spare sectors. Clusters marked as bad may be used to hide data.

• Unallocated/formatted data: Simple operating system tools can't see or extract data from a disk that is formatted or when some or all parts of the disk size are in an "unallocated space". Many forensic tools can extract data from unallocated spaces. However, disk repartitions can cause data or parts of the data to be corrupted or destroyed.

Why and How Much Technical Detail a Forensic Investigator Needs to Know?

In order to analyze a disk or find hidden data in that disk, it is necessary to know the layout of file systems and know which OS formatted the disk.

In this subject and most of the technical subjects in computer forensic investigations, investigators should have a certain level of technical knowledge and skills related to the investigation. In principle, investigators should have all required knowledge need to solve the case and find relevant evidences. The skills and levels of knowledge may vary from one case to another. It may also vary based on the forensic tools they are using. They need to know what tools to use and also use the right tools properly and effectively. Commercial and complex tools typically have more features. Some of those tools (such as EnCase, FTK, etc. will need more time to understand and master in comparison with small or lightweight tools.

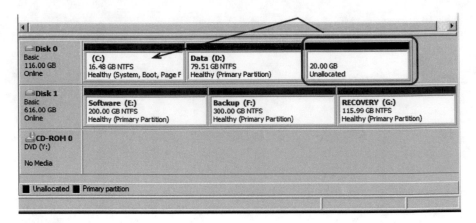

Fig. 10.9 An example of unallocated disk

When analyzing a suspect disk looking for evidences, it is important to understand the structure of this disk through its file system. It's also important to know which areas of this disk may have relevant data.

An investigator analysis a disk may need to deal with damages or corruptions in the file system. File systems can be corrected for natural reasons (e.g. disk bad sectors) or malwares or they can be possibly corrupted intentionally by hackers or suspects.

Fdisk and Format are two important commands in Windows operating system. Using Fdisk, you can create and change and display disk partitions. In Fdisk, you can decide the partition FAT type. In the same physical disk, two different logical drives or partitions can have different file systems. A logical space in the will be in an "unallocated" state (Fig. 10.9). Such disk partition will not be accessible to users. After creating the partition, the logical drive should be formatted before it can be used.

Disk Forensics in NTFS Systems

New Technology File System, NT FS or NTFS is a file system that is developed by Microsoft for Windows NT. NTFS considers some improvements over previous file systems FAT and High performance File System, OS/2 HPFS. Followings are main areas of improvements:

- **Security and encryption:** Unlike most other file systems, NTFS has encryption and built-in security features.
- **Metadata and Journaling:** NTFS log record metadata on disk changes. This feature that is added to NTFS can be very useful for system maintenance and forensic situations.
- **Advanced data structure and performance issues:**

- **Access Control Lists (ACLs):** ACLs are used in access control systems to control users' access on the different files.
- **Unicode support:** NTFS considers large encoding schemes to cover international languages (e.g. Chinese).
- **Compression:** NTFS has built-in compression methods to for scalability and size optimization.
- **Scalability and cluster size:** Disk addressing can take up to 2 ^ 32 clusters – 1. This is a very large possible cluster addressing space in comparison with most other file systems. This means that NTFS can deal with very large disk or file sizes. Unlike some other file systems, there will be no issue to partition a size with any possible size. A recent version of NTFS, NTFS5 can take up to 2 ^ 64 clusters – 1. NTFS uses a B-tree directory scheme to keep track of file clusters.
- **Multiple data streams:** In NTFS, a file can be associated with more than one application. From a forensic perspective, many malware examples use data streams to hide code.

In NTFS, a Master File Table (MFT) is a very important file in the file system. It keeps tracking of all files in the volume along with several metadata on those files. Such metadata is very important to check in any forensic investigations. Examples of such metadata include: creation date, entry modification date, accessed and last written date, the physical and logical sizes of the file in addition to permissions or security access for the file.

In the partition table in MBR, NTFS has a partition type number designation of 0x07 that you can look for to identify that this disk is partitioned/formatted using NTFS. Table 11.3 shows different indicators for different operating and file systems (Table 10.3).

Table 10.3 Examples of indicators for different operating and file systems

OS/FS indicator	FS/OS
0x00	Empty partition-table entry
0x01	DOS FAT12
0x02	XENIX /root file system
0x03	XENIX /usr file system
0x04	DOS FAT16 (up to 32 MB)
0x05	DOS 3.3+ extended partition
0x06	DOS 3.31+ FAT16 (over 32 MB)
0x07	OS/2 HPFS, windows NT NTFS, advanced Unix
0x08	OS/2 v1.0–1.3, AIX bootable partition, SplitDrive
0x09	AIX data partition
0x0A	OS/2 Boot Manager
0x0B	Windows 95+ FAT32
0x0C	Windows 95+ FAT32 (using LBA-mode INT 13 extensions)
0x0E	DOS FAT16 (over 32 MB, using INT 13 extensions)

Table 10.4 A sample of MFT metafiles

NO	Name	Description
0	$MFT	General MFT table to describe the positions, names, timestamps, etc.
1	$MFTMirr	A stored copy of the first four MFT record entries
2	$LogFile	Used to record changes for file system journaling.
3	$Volume	Contains volume object identifier (VOI), volume label, volume flags, and file system version.
4	$AttrDef	Specific attributes, data structure sizes, and other useful information.
5	.	Root directory.
6	$Bitmap	Select if a particular cluster on the volume is either used or free.
8	$BadClus	When bad clusters are detected, they will be allocated to this file.

Each MFT record number corresponds to one file number. Many tools support analyzing and querying files based on their MFT record number. For example, the the following command extracts the slack space of file with MFT number 40 and get the file size:. /istat -f ntfs /case1/image1 40. To extract the entire file its file slack:. / icat -sf ntfs /case1/image1 40 > /case1/file40.

MFT has also several metafiles that take the first several records. Table 10.4 shows examples of those metafiles.

Hackers/suspects can use techniques to hide data in bad clusters. The size of data that can be hidden with this technique is unlimited. Suspects can simply allocate more clusters to $BadClus (i.e. fake some sectors as bad sectors) and use it to hide data. You can check if clusters are allocated as to $Bad attribute of $BadClususing the command: istat /case1/image1 -f ntfs 8. The specific clusters, marked as bad can be further investigated using tools such as: dcat from Sleuth Kit. Tools for data carving (e.g. comeforth and foremost) can be also used to extract content from those clusters.

ADS (Alternate Data Stream)

ADS feature in NTFS has a significant impact in forensics. Users can hide files and data in files using ADS. If an MFT file record has more than 1 $DATA attribute, those additional $DATA attributes are called ADS (Alternate Data Stream). ADS can be used to hide data as it does not show up in directory listing and the file size of original file. ADS data hiding is relatively easy to accomplish and the size of data that can be hidden in ADS is unlimited.

To hide malecious.exe in an ADS called sample.txt type:

C:\ > echo malicious script text > sample.txt: malecious.exe

You can also do the same within a folder (Fig. 10.10).

```
C:\Users\Dell\Desktop>md folder1

C:\Users\Dell\Desktop>cd folder1

C:\Users\Dell\Desktop\folder1>echo Hide data in folder1>:hidden.txt

C:\Users\Dell\Desktop\folder1>dir
 Volume in drive C has no label.
 Volume Serial Number is 58B2-59B8

 Directory of C:\Users\Dell\Desktop\folder1

04/21/2017  04:40 PM    <DIR>          .
04/21/2017  04:40 PM    <DIR>          ..
               0 File(s)              0 bytes
               2 Dir(s)  318,847,262,720 bytes free

C:\Users\Dell\Desktop\folder1>
```

Fig. 10.10 An example of hiding data using ADS

```
C:\Users\Dell\Downloads\ADSCheck-v1.0.0.46-binary>ADSCheck.exe C:\Users\Dell\Desktop\folder1
ADSCheck v1.0.0.46. Copyright(C) 2005, Dr. Hatem Kawashti.
ADSCheck comes with ABSOLUTELY NO WARRANTY; This is free software, and you are
welcome to modify, copy and redistribute it under certain conditions;
type 'ADSCheck /L' for license details.

Found 1 ADS in file: C:\Users\Dell\Desktop\folder1 .. listing size, name:-
      22       hidden.txt

Done. 1 ADS found.  1 Files in 1 Directories processed.

C:\Users\Dell\Downloads\ADSCheck-v1.0.0.46-binary>
```

Fig. 10.11 An example of using ADSCheck to detect ADS

ADS Detection Tools

ADS is not typically discovered using normal Windows operating system tools. ADS cannot be detected with the majority of file browsers such as Windows Explorer or the DOS command "DIR". There are legitimate uses of ADS and we should not assume that every single ADS encountered is used to hide data.

Several special tools can be used to detect ADS. Many large commercial forensic tools (e.g. autopsy/TSK, FTK, etc.) can analyze ADS. Here are other examples of those currently available free/open source tools:

- CAT tool from Nix Utilities: https://sourceforge.net/projects/unxutils. The following command can be used to retrieve/recover the content of the hidden file (assuming that we know the hidden file:

```
C:\Users\Dell\Downloads\UnxUtils\usr\local\wbin>cat "C:\Users\Dell\Desktop\folder1:hidden.txt">"Recovered.txt"
```

- ADSCheck (https://sourceforge.net/projects/adscheck). Figure 10.11 shows a simple usage example.
- StreamArmor: Fig. 10.12 shows a simple usage example.
- ADSSpy (Fig. 10.13)

Fig. 10.12 An example of using StreamArmor to detect ADS

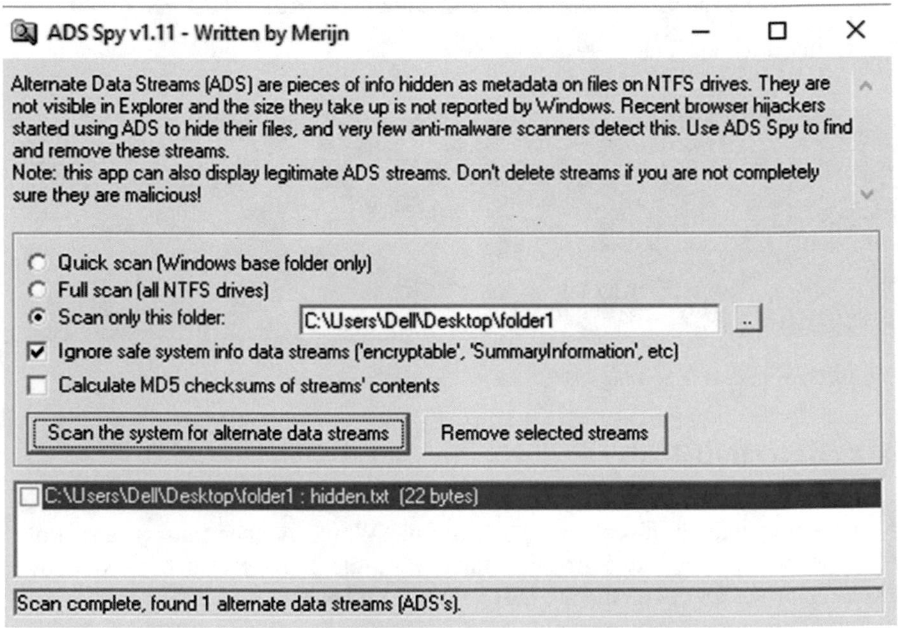

Fig. 10.13 An example of using ADSSpy to detect ADS

Disk Forensics in Ext. Systems

Extended File System (Ext) is a File system (family) introduced with Linux and consists of Ext, Ext2, Ext3, Ext4. Ext file systems share many properties with NTFS: Access Control Lists (ACLs), fragments, undeletion and compression and journaling (beyond Ext2). An Ext file system starts with a superblock located at byte offset 1024 from the volume start (Block 0 or 1). Figure 10.14 shows a high level layout of Ext partition.

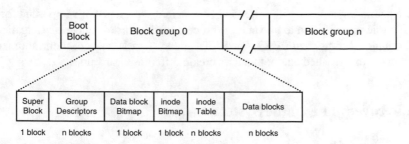

Fig. 10.14 An Ext file system high level layout

The first version (Ext) came to solve issues with Minix or Mint previous file systems. The two main issues Ext targets were extending maximum partition size up to 4 TB (Ext has a maximum partition size of 2 GB) and also the file name length to 14 characters. Soon Ext2 replaces Ext due to many problems in Ext file system and Ext2 became the standard for Linux operating systems. The main problem with Ext2 which leads to introducing Ext3 is that Ext2 does not support journaling. As it has no journaling, Ext2 is still preferred on USB flash drives as it requires fewer write operations.

In Ext file systems, every file or directory is represented by an inode (index node). The inode includes metadata about the size, permission, ownership, and location on physical disk of the file or directory, number of blocks used, access time, change time, modification time, deletion time, number of links, fragments. For forensic analysis, although they don't hold contents, inodes are the primary repository of metadata.

Understanding the journaling process in file systems can help forensic investigators recover deleted or corrupted files if necessary. Journaling is a process used in file systems for crash recovery situations. File system archives activities in normal situations so that when a problem occurs, it can recover to a safe situation. The journal process works by caching some or all of the data writes in a reserved part of the disk before they are committed to the file system.

In Ext3, the journal is a fixed size log which regularly overwrites itself. The journaling process can be accomplished using different approaches or algorithms (e.g. Journal, ordered and write-back) in the different operating systems. The first factor that impacts the journaling process is the performance or how much impact the journaling process can cause to the operating system and normal operations. This takes the form of efficiency or resources' consumption as well as speed. The second factor that impacts or decides the journaling process is the amount of data to journal. There is a clear trade-off between performance and amount of data or metadata to log. From a forensic point of view journal mode will be better than ordered or write-back as it offers metadata and recovery option unlike the other two options that can only help in recovery (without significant metadata). The default option in Ext3 is "ordered" and hence little metadata can be extracted from investigated file system activities.

The journaling process will write/modify data in disks. Hence, it's important for a disk forensic investigation to prevent this process while a disk is under investigation. For example, if a disk with EXT3 or EXT4 file system is hashed, then mounted/un-mounted then hashed again, this may create different hash values.

File Recovery in Ext File Systems

Similar to most file systems in Ext file systems, deleting a file does not erase file content but only the file system link or record of the file. For the file system, once a file is deleted, its space will be claimed unallocated and any new system activity (e.g. creating new files, installing applications, operating system updates, or journaling) can *possibly* use this unclaimed space.

If files are deleted and their space is not yet used by other files or applications, the process of retrieving such data or files is relatively easy and can be provided by a large number of tools from operating systems, forensic tools, free, open source, etc. However, the process can be very complex if this space is used partially or completely by other files or applications. Followings are steps that can be followed in a forensic investigation to recover a file or some files:

- Identify their inode numbers. In Ext file systems files are uniquely linked with different inode numbers. Several tools can be tried to extract Ext file system inode numbers (e.g. TSK ils, ls, stat, istat, df, find, etc. The command (ls –i) shown in Fig. 10.15 lists the inode numbers for the files/folders.

  The inodes do not contain the content of the files but their metadata. Figure 10.16 shows using the command (ls –li) on a specific file.

```
% ls -i
2637825 bin        983041 etc         1572865 lib        2981889 media  2531329 root     106497 selinux     81921 usr
 196609 boot             2 home        1761281 lib64       2129921 mnt       6416 run     2457601 srv        425985 var
```

Fig. 10.15 The command (ls –i) to list inode numbers

Fig. 10.16 Using the command (ls –li) and inode details

```
# ls -li file1
918827 -rw-r--r-- 1 root root 19 2016-12-05 11:08 file1
# istat /dev/mapper/home 918817
inode: 918827
Allocated
Group: 112
Generation Id: 3173542730
uid / gid: 0 / 0
mode: rrw-r--r--
Flags:
size: 0
num of links: 1

Inode Times:
Accessed:           Sun Dec  5 11:08:49 2016
File Modified:      Sun Dec  5 11:08:49 2016
Inode Modified:     Sun Dec  5 11:08:49 2016

Direct Blocks:
```

- Using commands such as stat, istat or find, you can make queries on specific inodes to get their status and metadata. (e.g. find. -inum 434,404 –print).
- Knowing in which (block group) the subject inode exist, can help us search for more details about the block that the inode belong to. For example, we can use a command such as: fsstat to make queries on block groups. Block groups are described in a block or set of blocks called "group descriptor table".
- From the journal (if supported), inodes can be extracted using commands such as dumpe2fs, jcat, dcat, blkcat, etc.

Disk Forensics in HFS+ Systems

Hierarchical File Systems (HFS and HFS+ or extended HFS) are file systems used by Apple in their operating systems; MAC as successors to earlier MAC file system; MFS. HFS replaces MFS to solve problems related with the evolution of disk sizes and the need to deal with large disk sizes, file sizes and names, etc. Similar to NTFS, HFS replaced MFS flat table structure with the Catalog File which uses a B-tree structure that could be searched very quickly even when size is large. HFS+ is introduced in MAC OS 8.1 with better performance and internationalization/encoding features in comparison with HFS.

The Trash is represented on the file system as a hidden folder,. Trash, on the root directory of the file system (Fig. 10.17). However, similar to Windows, Trash folder content can be retrieved using the mouse or command line. HFS+ preserves dates and time stamps when moving files to and from the Trash. Similar to other file systems, deleted files can be retrieved if their place is not overwritten.

Table 10.5 shows basic MAC disk and partition commands

Figure 10.18 shows the result of running Sleuth Kit's mmls command on an image with HFS.

Fig. 10.17 MAC hidden trash folder

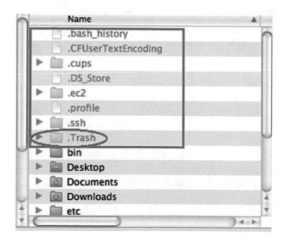

Table 10.5 MAC basic disk commands

diskutil list	List Connected Disks
diskutil info <disk>	Disk information (use: /dev/disk#, disk#, or partition /dev/disk#s#)
pdisk –l /dev/disk3	List partitions using Apple partition Map Format
gpt –r show [l]	List partitions using GUID partition (-l to show label rather than GUID)
Mmls <diskimage>	Display partitions using The Sleuth Kit
hdiutil imageinfo *.dmg	Disk Image Information including partition data
hdiiutil fsid *.mg	Volume header information of disk image

```
sudo mmls /dev/disk1
UID Partition Table (EFI)
Offset Sector: 0
Units are in 512-byte sectors

       Slot     Start        End          Length       Description
000:   Meta     0000000000   0000000000   0000000001   Safety Table
001:   -------  0000000000   0000000039   0000000040   Unallocated
002:   Meta     0000000001   0000000001   0000000001   GPT Header
003:   Meta     0000000002   0000000033   0000000032   Partition Table
004:   000      0000000040   0000409639   0000409600   EFI System Partition
005:   001      0000409640   0030703575   0030293936   HFSImage
006:   -------  0030703576   0030965759   0000262184   Unallocated
```

Fig. 10.18 The result of (mmls) command on an image with HFS

```
          0  1  2  3  4  5  6  7  8  9  A  B  C  D  E  F   0123456789ABCDEF
0400h:   48 2B 00 04 80 00 21 00 48 46 53 4A 00 00 00 75   H+..€.!.HFSJ...u
0410h:   D3 B9 00 BF D3 B9 63 60 00 00 00 00 D3 B9 63 2F   Ó¹.¿Ó¹c`....Ó¹c/
0420h:   00 00 00 42 00 00 00 13 00 00 10 00 00 39 C7 F6   ...B........9Çö
0430h:   00 39 A3 A8 00 03 80 1D 00 01 00 00 00 01 00 00   .9£"..€.........
0440h:   00 00 00 6D 00 00 00 CB 00 00 00 00 00 00 00 01   ...m...Ë........
0450h:   00 00 00 00 00 00 00 00 00 00 00 00 00 00 00 00   ................
0460h:   00 00 00 00 00 00 00 00 37 A4 E9 FD EA FC 14 EA   ........7¤éýêü.ê
```

Fig. 10.19 HFS+ volume header

Volume Headers

HFS+ volumes can be distinguished through their volume headers. The volume provides important details such as: creation, modification and checked dates. In addition to the volume header, there is also the alternate volume header. The volume header is located at offset 0x400h, 1024 bytes from the start of the volume (Fig. 10.19).

Memory Forensics

Different computing systems (e.g. desktops, laptops, tablets, smart phones, routers, etc) include memories. Unlike disk drives, memories include only most recent running applications and activities. In comparison with disk data, memory includes the dynamic portion of the data. Usually, the most recent activities are very important for forensic investigations when computing systems are part of possible evidence sources.

In comparison with analyzing disk data, memory data is harder to analyze given that it is typically in binary or hexadecimal, not ASCII human readable format. Special tools such as hexadecimal editors are necessary to analyze memory data. In this section we will evaluate different environments and tools to analyze memory data.

Memory can be read only (ROM) or erasable random access (RAM). BIOS and firmware are examples of ROM that will not be lost when restarting the machine. While ROM may have sometimes relevant forensic data, however, largely possible forensic data will be found in RAM. Our focus in this section is in RAM.

When users start software applications, data/code that belong to those applications are brought to the memory. The acquisition and the analysis of RAM data is harder than static disk data. Users can easily erase RAM data by shutting down or restarting their computing systems. Forensic investigators may also unintentionally erase or tamper RAM data if they did not follow proper procedures. As part of the integrity of any source evidence, investigators need to prove that they did not insert/edit any footprints in the source evidence.

The size of the memory is limited and typically much smaller than the disk size. Computing systems include different types of memory management algorithms to make best judgment on what data to stay in the physical memory and what data to swap out of the physical memory (e.g. to the virtual memory).

The virtual memory is a supporting memory size that can be taken from the disk size to support the physical memory (Fig. 10.20). System can then swap pages of data between physical and virtual memory. In memory forensics, virtual memory should be extracted and dealt with as a memory rather than disk data.

Forensic Relevant Information That Can Be Found in Memory

When analyzing memory, a forensic investigator should first study the case and understand all its details. This can help her/him know what information to be looking for. On the other hand, most commercial memory analysis tools prepare a list of pre-defined categories of information that can be relevant in general. Here is a list of such categories of information:

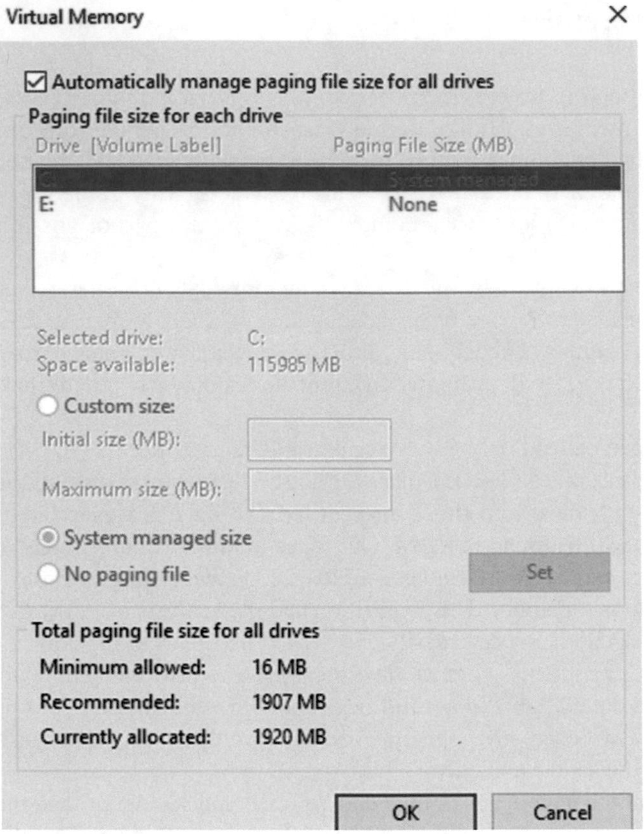

Virtual Memory   ✕

☑ Automatically manage paging file size for all drives

Paging file size for each drive

Drive [Volume Label]	Paging File Size (MB)
C:	System managed
E:	None

Selected drive: C:
Space available: 115985 MB

○ Custom size:
Initial size (MB):

Maximum size (MB):

◉ System managed size
○ No paging file   Set

Total paging file size for all drives

Minimum allowed: 16 MB
Recommended: 1907 MB
Currently allocated: 1920 MB

OK   Cancel

Fig. 10.20 Virtual memory settings

- User credentials and account details: When users login to their accounts in emails, social networks, websites, etc. their credentials can be stored and extracted from the memory.
- Most recent opened applications: In many computer crime cases, it is important to know the most recent applications the victim or the suspect where using especially if this is synchronized with the attack or the crime.
- Most recent opened or accessed data or files. This including: created, modified or accessed files.
- Content of email messages, posts or comments in social networks, pictures, videos, etc. that were recently created, modified or accessed.
- Most recent network connections, visited websites, etc.
- Malware analysis depends significantly on memory analysis. Many malwares reside in memory and may not write themselves in disks. Behavior and dynamic analysis of malwares are conducted for either detection or analysis.

Memory Forensic Tools

Many simple or generic tools such as tools to search for strings (e.g. strings, grep, etc.), hexadecimal editors can be used to extract data from memory. Operating systems built-in tools such as debugging tools can be also used to analyze memory data. For example, Unix debugger gdb can be used to analyze the extracted raw memory images. Memory forensic tools can be largely categorized based on the general function or stage into the two following categories. Some of the listed tools can perform both functions and tasks.

1. Memory dumping tools: Tools are needed to extract current memory data or status from current computing system. Most of those tools are console-based to extract memory live from running systems. Examples of those tools (dumpit (Fig. 10.21),fmem, LiME, Second Look, mdd, kntdd, etc.). A comprehensive list of those tools exist in the link: http://forensicswiki.org/wiki/Tools: Memory_Imaging.

   Forensic commercial tools such as Accessdata FTK and EnCase provide also options to capture memory (Fig. 10.22).

   There are several examples of hardware-based memory acquisition tools. In one example, CaptureGUARD (http://www.windowsscope.com), the tool is a PCI card that can be connected to a computer to extract memory. This involves suspending computer processor and using Direct Memory Access (DMA) to get a memory copy.

2. Memory analysis tools: Once memory dump from subject source is acquired, the next stage is to analyze the memory dump looking for case relative evidence or information. As we described previously, most of the memory forensic tools have predefined artifacts that the tools will be looking for in the subject evidence.

   - Volatility: A popular open source memory forensic tool or collection of tools: (https://github.com/volatilityfoundation/volatility, http://www.volatility-foundation.org/). The tool is used to extract several memory-related artifacts (Fig. 10.23).

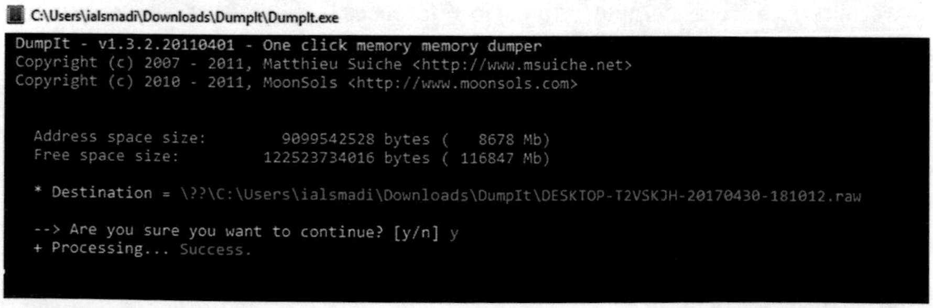

```
C:\Users\ialsmadi\Downloads\DumpIt\DumpIt.exe

DumpIt - v1.3.2.20110401 - One click memory memory dumper
Copyright (c) 2007 - 2011, Matthieu Suiche <http://www.msuiche.net>
Copyright (c) 2010 - 2011, MoonSols <http://www.moonsols.com>


  Address space size:        9099542528 bytes (    8678 Mb)
  Free space size:         122523734016 bytes (  116847 Mb)

  * Destination = \??\C:\Users\ialsmadi\Downloads\DumpIt\DESKTOP-T2VSKJH-20170430-181012.raw

  --> Are you sure you want to continue? [y/n] y
  + Processing... Success.
```

Fig. 10.21 Dumpit memory dump tool

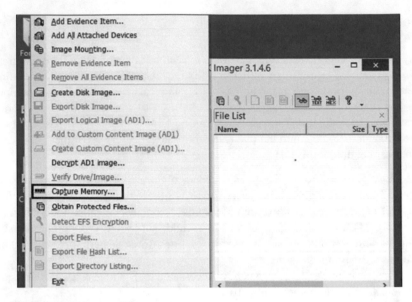

Fig. 10.22 FTK Imager memory capture

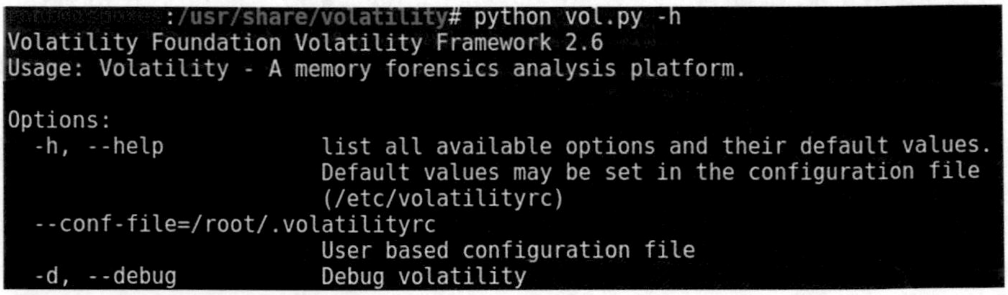

```
                 :/usr/share/volatility# python vol.py -h
Volatility Foundation Volatility Framework 2.6
Usage: Volatility - A memory forensics analysis platform.

Options:
  -h, --help               list all available options and their default values.
                           Default values may be set in the configuration file
                           (/etc/volatilityrc)
  --conf-file=/root/.volatilityrc
                           User based configuration file
  -d, --debug              Debug volatility
```

Fig. 10.23 Starting Volatility tool in Kali

Volatility framework is very popular. It is used in many security training courses. Several computer forensic references indicate that the framework has been effectively used to discover interesting evidences in many forensic cases.

- Redline: This memory forensic tool can be used for Windows environment (https://www.fireeye.com).
- Volatilitux: : A Linux version of Volatility. It supports different architectures for memory dumps: ARM, x86.

Figure 10.24 shows a list of Linux-based memory analysis tools provided by Forensicswiki website.

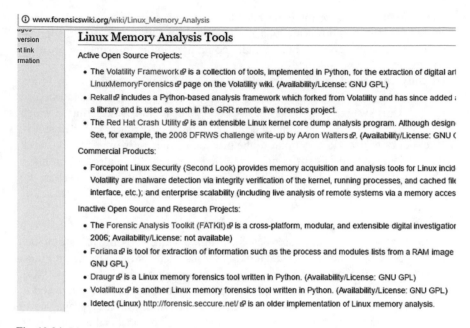

① www.forensicswiki.org/wiki/Linux_Memory_Analysis

Linux Memory Analysis Tools

Active Open Source Projects:

- The Volatility Framework ⌕ is a collection of tools, implemented in Python, for the extraction of digital art
  LinuxMemoryForensics ⌕ page on the Volatility wiki. (Availability/License: GNU GPL)
- Rekall ⌕ includes a Python-based analysis framework which forked from Volatility and has since added ;
  a library and is used as such in the GRR remote live forensics project.
- The Red Hat Crash Utility ⌕ is an extensible Linux kernel core dump analysis program. Although design
  See, for example, the 2008 DFRWS challenge write-up by AAron Walters ⌕. (Availability/License: GNU (

Commercial Products:

- Forcepoint Linux Security (Second Look) provides memory acquisition and analysis tools for Linux incid
  Volatility are malware detection via integrity verification of the kernel, running processes, and cached fil
  interface, etc.); and enterprise scalability (including live analysis of remote systems via a memory acces

Inactive Open Source and Research Projects:

- The Forensic Analysis Toolkit (FATKit) ⌕ is a cross-platform, modular, and extensible digital investigatior
  2006; Availability/License: not available)
- Foriana ⌕ is tool for extraction of information such as the process and modules lists from a RAM image
  GNU GPL)
- Draugr ⌕ is a Linux memory forensics tool written in Python. (Availability/License: GNU GPL)
- Volatilitux ⌕ is another Linux memory forensics tool written in Python. (Availability/License: GNU GPL)
- Idetect (Linux) http://forensic.seccure.net/ ⌕ is an older implementation of Linux memory analysis.

Fig. 10.24 Linux memory analysis tools

Forensic Investigations in Windows Operating Systems

Regardless of the operating system under investigation, there are some major generic activities that should be conducted in any operating system:

- **Review all system logs:** Each operating system keeps records of different activities that occur in the operating system. They can show details about system users, installation and used applications, and the different types of activities accomplished on those systems. Regardless of the nature of the forensic case understudy, it is always important to investigate all system logs. It is also important to verify the integrity of those logs as some experience suspects or hackers can tamper those logs to intentionally mislead investigations.
- **Perform keyword searches:** This is a task that is highly dependent on the nature of the forensic case. Forensic investigators should read thoroughly the details of the case in order to be able to propose basic keywords to start their search with. The results of those initial keywords can trigger further searches. Many commercial tools have a list of structured and predefined keywords (e.g. email addresses, physical addresses, phone numbers, IP addresses, etc.) that can be used.
- **Review relevant files:** What can decide that a particular file or directory is relevant or not is of course the case and the nature or details of the case. However, we need to know the certain folders or directories that such information

can be found in. For example, we need to search in addition to system logs and directories, Internet browsing logs or history, last installed or accessed applications and files and so on. Those usually contain valuable information for most forensic cases.

- **Identify unauthorized user accounts or groups:** For the integrity of any forensic case, it is important to know in any investigated system, all the users who had access to this system. In some cases, suspects can be themselves victims in which their systems were used to commit crimes without their explicit knowledge.
- **Backdoors or Rootkits**: Similar to the previous issues, it is important to check that an investigated system is not controlled remotely. It is also important to check that no tasks are auto-scheduled to work (in some date or time). There are standards tools that can be used to test for such instances (Check chapter skills' section).
- **Identify malicious processes:** In malware analysis, which is considered as a major function in digital forensics, investigators are looking for suspicious behaviors and programs. They are looking for instances of those rogue programs and they are also interested to understand and analyze the behavior of those rogue or malicious processes or programs. For example, they are interested to know how they intrude or access victim machines and how they propagate to other machines. They want also to know their payload or what is the goal of such malware?

Since it started with DOS operating system, Microsoft launched several operating systems including: Windows 95, 98, XP, NT, 2000, ME, 2003 server, Vista, 7, 8, 2008 server, 2012 server, 2016 server and Windows 10. Forensic investigators may have to deal with images coming from anyone of those different operating systems. Even as Microsoft stopped supporting most of those operating systems, forensic investigators will have to study many technical details about them.

Windows operating systems are very popular. Many forensic cases included the analysis of different versions of Window operating systems. Forensic analysts still practice and use some tools for early Windows operating systems (e.g. DOS, Windows for work groups, 95 and 98). On the other hand, Microsoft stopped supporting all those old operating systems. In comparison with forensic tools and support for other operating systems, tools and support for forensic activities in Windows operating system is large in the market. Users can find many free, open source or commercial tools to conduct different types of forensic activities in Windows operating systems.

We discussed disk forensics in Windows file systems (i.e. FAT and NTFS) in earlier sections. In addition to disk forensics, operating systems can have many artifacts for forensic interests. We will describe some of those that are most relevant to Windows operating systems.

In Windows environments, there are common forensic artifacts that can be searched or investigated in any suspect Windows image. Table 10.6 shows a list of those major Windows artifacts and their location in recent Windows operating systems.

Table 10.6 An example of important forensic artifacts in Windows

Artifact	Typical location
Registry and User-assist keys: This includes information about users, current users, sessions, installed applications, etc.	HKEY_LOCAL_MACHINE \SYSTEM: \system32\config\system HKEY_LOCAL_MACHINE \SAM: \system32\config\sam HKEY_LOCAL_MACHINE \SECURITY: \system32\config\security HKEY_LOCAL_MACHINE \SOFTWARE: \system32\config\software HKEY_USERS \UserProfile: \winnt\profiles\username HKEY_USERS.DEFAULT: \system32\config\default HKEY_CURRENT_CONFIG
Event Logs	C:\Windows\System32\config
Internet Browser artifacts	Browser directory
Volume shadows	Settings-Computer and disk management
File systems	Settings-Computer and disk management
Link and recent files	AppData\Roaming\Microsoft\Windows\Recent
Cookies	Several different locations

Windows Registry

The first and most important artifact to investigate is the registry. Windows registry is a hierarchical database with links to different files and activities. Registry is organized into Hives, Keys and sub-keys that contain relevant values. The six hives (System, SAM, Security, Software, Users, User_Default) are listed with their names and locations in Table 11.6. Hives are the files stored in the disk while the keys are the logical registry roots.

Examples of forensic relevant information that can be extracted from Registry include: Computer Name, Operating System installation dates, Software programs installed & uninstalled, USB devices connected, most recently created/used/accessed files, IP addresses, Network shared information, malware activity and information about the user level activity. Figure 10.25 shows main screen of Windows 10 registry.

Internet Traces

We will have a chapter in this book dedicated to "web forensics" and the different types of forensic artifacts that can be extracted from the web or from the users' Internet usage and traces.

The first important types of applications related to Internet usage that should be investigated are the web browsers. Web browsers are software applications that enable users to access different websites through the internet. History of those browsers can help us see which websites subject user was visiting, which pages in those websites, when they were visiting those websites and many other important information. Of course the value and important of such information can vary from one forensic case to another.

Fig. 10.25 Windows 10
Hives

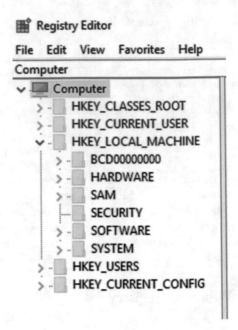

From OSI layers' perspective, most of the information we can extract from web browsers can fall within layer 7 (i.e. application layer). In comparison to network analysis and forensic tools (e.g. Wireshark) that typically detect information in lower layers (e.g. L2-L3), information that can be extracted from browsers' history is easier to use and understand by humans or investigators without the need for significant skills or special high-end forensic tools. We will describe briefly how to search for Internet usage history in Windows most 3 popular browsers: MS internet explorer, Google Chrome and Firefox.

- MS Internet explorer

Microsoft has their own browser that will be installed with Windows operating systems. The browser options (Fig. 10.26) can be used to view or set the location where browsing history is stored.

Users can also make/change settings related to the size of the history or cache or for how long to keep such cache. Users can opt not to store browsing history. Internet browsing history can be used to accelerate bringing pages from the Internet where some pages can be brought from the cache if they exist without the need to collect them from the Internet. Users can also clear their browsing history. In such cases, forensic investigators can search for pointers for the Internet history (e.g. cookies, recent files or shortcuts). Those can show indications of visited websites, watched videos, pictures, files, etc. while they may not help in retrieving those artifacts.

- Google Chrome

Google has two main versions of browsers: Google chrome and Chromium. Chromium is recently introduced to target Desktop environments. There are some

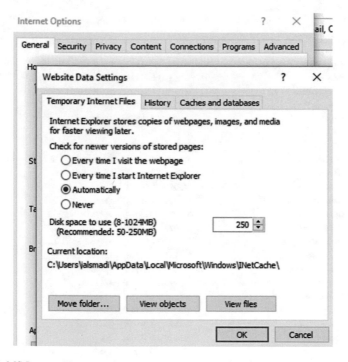

Fig. 10.26 MS Internet history settings

differences between the two versions on where the default location of their Internet browsing history. In most recent Windows operating systems, this default location is in user application data folder (e.g. C:\Users\%USERNAME%\AppData\Local\ Google\Chrome\User Data\Default\Preferences). Configuration can be seen or changed through access Google Chrome configuration menus. In addition to cache or history content, the folder can contain many artifacts that can be useful for forensic investigations (Fig. 10.27).

One of the main file to investigate in this folder is the (History file). This history file uses SQLite database format (Fig. 10.28).

Different time stamps can be extracted about the user's visits to the different websites, time and duration of such visits.

Another important file in the same application data folder to investigate is the (Cookies) file. Similar to the (History) file, Cookies file uses SQLite database format.

- Mozilla Firefox

Firefox browser is a free and open source browser developed by Mozilla foundation. Some of the distinguished features in Firefox in comparison with other browsers are: The support of a large number of add-ons, anonymous browsing, etc. Similar to other browsers, the default location of forensic artifacts in most recent windows operating systems is the application data folder (e.g. C:\Users\%USERNAME%\ AppData\Roaming\Mozilla\Firefox\Profiles\%PROFILE%.default\places.sqlite).

Name ^	Date modified	Type	Size
Application Cache	4/30/2017 11:57 AM	File folder	
Cache	5/12/2017 9:35 PM	File folder	
data_reduction_proxy_leveldb	5/12/2017 6:19 PM	File folder	
databases	5/12/2017 6:43 PM	File folder	
Extension Rules	5/10/2017 10:32 PM	File folder	
Extension State	5/12/2017 6:19 PM	File folder	
Extensions	5/11/2017 10:56 AM	File folder	
File System	5/12/2017 6:43 PM	File folder	
GCM Store	5/12/2017 6:19 PM	File folder	
GPUCache	5/3/2017 1:22 PM	File folder	
IndexedDB	5/12/2017 7:21 PM	File folder	
JumpListIcons	5/12/2017 9:50 PM	File folder	
JumpListIconsOld	5/12/2017 9:45 PM	File folder	
Local Storage	5/12/2017 9:35 PM	File folder	
Media Cache	5/12/2017 7:02 PM	File folder	
Platform Notifications	5/12/2017 7:11 PM	File folder	
Service Worker	5/2/2017 12:41 PM	File folder	
Session Storage	5/12/2017 9:35 PM	File folder	
Sync Data	4/30/2017 11:52 AM	File folder	
Thumbnails	5/12/2017 6:19 PM	File folder	
Bookmarks	5/12/2017 6:21 PM	File	10 KB
Bookmarks.bak	5/12/2017 4:35 PM	BAK File	10 KB
Cookies	5/12/2017 9:51 PM	File	3,168 KB
Cookies-journal	5/12/2017 9:51 PM	File	0 KB
Current Session	5/12/2017 9:36 PM	File	1 KB

Fig. 10.27   Google chrome forensic artifacts

History C:\Users\ialsmadi\AppData\Local\chromium\User D... Type: File	Date modified: 5/12/2017 9:50 PM Size: 21.6 MB
History-journal C:\Users\ialsmadi\AppData\Local\chromium\User D... Type: File	Date modified: 5/12/2017 9:50 PM Size: 0 bytes
History Provider Cache C:\Users\ialsmadi\AppData\Local\chromium\User D... Type: File	Date modified: 5/7/2017 6:10 PM Size: 2.76 MB

Fig. 10.28   Google chrome history file

Event Viewer

Event viewers in Windows operating systems include records and logs of many events that occur in the operating system and installed applications. While they meant mainly for maintenance and troubleshooting purposes, yet they can be very valuable for forensic investigations (Fig. 10.29).

Event viewers include three major categories of loggings: System (i.e. operating system logging activities), applications and security loggings. Logged events in those categories can be in one of three types: Error, warning or information. Each event includes information on when the event occurs, through which user, application, etc. Users can also create customized event logs in the applications they use or develop. Events can be exported as text files and many tools can be used to extract artifacts from those events.

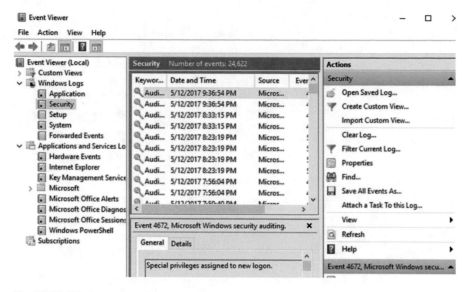

Fig. 10.29 Windows event viewer

Web Logs

If inspected machine or image has a web server (e.g. Microsoft Internet Information Server IIS, or Apache (https://www.apache.org/)), web logs can be extracted from the machine. Such web logs may include many forensic artifacts related to the users or visitors of this web site and their using activities or behaviors. More details will be provided in Web forensics chapter.

Forensic Investigations in Linux Operating Systems

As an open source operating system, Linux includes many versions/flavors deployed in different environments. We described earlier Linux-based disk forensics, image acquisition and analysis. Table 10.7 shows major directories in standard Linux distributions.

In comparison with Windows, Linux operating systems do not have a registry hierarchical data structure or centralized registry. Those registry traces exist in different directories. System or machine wide settings and software are located in the folder (/etc). Users' settings are typically located as hidden files in the users' home directory. Kernel related parts exist in /proc. folder. Different desktop environments have their own registry APIs: Gconf, dconf/GSettings in GNOME and Xfconf in Xfce.

Forensic investigators should also get themselves with important system files and their locations. We showed in Table 10.8 a sample of those system files that can be significant for most forensic analyses. Users should find comprehensive lists in Linux manuals, books or online references.

Table 10.7 Major directories in standard Linux distributions

Linux directory	Details
root directory	This is the base of the file system's tree structure.
−/bin	Binary files for the OS
−/dev	The device files
−/sbin	System administrative binaries
−/home	Conventional location for users' home directories.
−/etc.:	system configuration files
lost + found	Storage for recovered files

Table 10.8 A sample of Linux forensic relevant system files

File	Details
/dev/had	First IDE hard drive on the system
/etc/aliases	Contains aliases used by sendmail and other mail transport agents.
/etc/bashrc	Contains global defaults used by the bash shell
$HOME/.bash_history	Command history
/etc/exports	Contains file systems available to other systems on the network via NFS.
/etc/fstab	The file system table contains the description of what disk devices are available at what mount points.
/etc/shadow	Hashed (e.g. MD5) versions of passwords
/etc/group	Holds information regarding security group definitions.
/etc/grub.conf	Grub boot loader configuration file
/etc/hosts	Contains host names/IP addresses used for name resolution in case a DNS server is unavailable
/etc/mtab	Information about currently mounted devices and partitions
/etc/sudoers	Usually, shows users with admin privileges
/etc/passwd	Contains information about registered system users.
/etc/resolv.conf	Domain name servers (DNS) used by the local machine
/proc/cpuinfo	Contains CPU related information
/proc/filesystems	Contains information about file systems that are currently in use
/proc/ioports	A list of I/O addresses used by devices connected to the server
/proc/meminfo	Contains memory usage information for both physical memory and swap
/proc/modules	Lists currently loaded kernels
/proc/mounts	Displays currently mounted file systems
/proc/stat	Contains various statistics about the system
/proc/swaps	Contains swap file utilization information
/proc/version	Contains Linux version information
/var/log/lastlog	Stores information about the last boot process
/var/log/messages	Contains messages produced by the syslog during the boot process
/var/log/wtmp	A binary data file holding login time and duration for current users

Many of the forensic tools we described earlier can be used for Windows as well as Linux. For example, most commercial forensic tools (e.g. FTK, Sleuth kit) have supported versions for the different operating systems. We also described version that for Volatility open source memory analysis framework, a special version (Volatilitux: https://code.google.com/p/volatilitux) support Linux.

To check integrities of applications installed in Linux, hashing can be used to make sure that installed applications are original not tampered by hackers. We explained earlier how to create a local hashing server based on an open source application and NSRL/Hash databases (https://www.nsrl.nist.gov/Downloads.htm). This hashing server can be used to check the integrity of all Linux installed applications (installed and system applications, shared libraries, etc). Earlier Table, 10.8 shows location of Linux logon log files. In addition to the three mentioned earlier in the Table (i.e. lastlog, messages, wtmp), there are many interesting login log files in (/var./log) directory. Student is encouraged to collect more details about those log files and what information can be extracted from them. They can be all reached using (/var./log) and then one of the following names: messages, dmesg, auth.log, boot.log, daemon.log, dpkg.log, kern.log, lastlog, mail.log, user.log, Xorg.x.log, alternatives, btmp, cups, anaconda.log, yum, cron, secure, wtmp, utmp, failog, httpd, apache2, lighttpd, conman, mail, prelink, audit, setroubleshoot, samba, sa, and sssd. In addition to those log files, there are some logs in (/var./admn) directory that can be only seen by admins.

Using the console to conduct some basic forensic steps can be hard especially for users who are not familiar with such console commands. For example, a forensic investigator who wants to find hidden files/folder in Windows environments can use the mouse to unhide files and folders. On the other hand, in Linux, they need to craft a special Console command (e.g. find. –type d –name ".*" –print0 | cat –a) to do so. As such, conducting forensic analysis in Linux environments can take more time and require more technical skills. On the other hands, some native command line tools in Linux can perform important forensic tasks. For example, you can use netstat to search for open ports and associated applications (netstat –anp). In order to do that in Windows, possibly the easy way is to install a port scanner tool (e.g. nmap). Similarly, you can use a command line tool (lsof: List open files) to list applications or processes with their open files, ports, etc.

The search through the (/tmp) folder can be also important for forensic analysis in Linux. It can show recent installed or accessed programs, intrusion traces, etc. typically installing an application or executable requires many intermittent activities that can leave traces in the (/tmp) folder or directory.

Different web browsers can be used in Linux environments. The most popular are: Firefox and Google Chrome. History or data from those browsers can exist in the directories:

- Firefox: $HOME/.mozilla/firefox/*.default
- Chrome: $HOME/.config/chromium/Default

One of the popular open source security operating systems is Kali Linux (www. kali.org). You can install Kali in Forensic live mode or as a virtual image. Many Linux forensic tools come pre-installed in Kali: http://docs.kali.org/general-use/ kali-linux-forensics-mode.

Forensic Investigations in MAC Operating Systems

Similar to Linux, Apple MAC operating system is evolved from UNIX. Most of major forensic artifacts are similar in the different operating system, while their locations can be different. Apple MAC artifacts and locations can be different based on the OS platform (i.e. Desktops, Laptops, Tablets or smart phones). Newer MAC operating systems are distinguished by the letter (X) from older version. The mobile OS is distinguished by the letter (i; iOS).

Apple is continuously evolving their operating systems especially in terms of security architecture or framework. We will describe more details in this in Mobile forensics chapter.

The followings are locations from MAC on mobile devices.

- **System version:** /System/Library/CoreServices/SystemVersion.plist
- **User preferences:** In MAC, typically they exist in: %%users.homedir%%/Library/Preferences/*. Many artifacts exist in this directory including preferences for: Global, iCloud, sidebar lists, dock, attached iDevices, quarantine event database, etc.
- Different files and metadata can be found in the backup directory.
- **Call History**: %%users.homedir%%//Library/CallHistory
- **Text messages**: %%users.homedir%%//Library/SMS
- **Address Book**: The location of this file is %%users.homedir%%//Library/AddressBook. There are primarily 2 databases in AddressBook: sqlitedb: Contains contact information and sqlitedb: Contains contact images. They use the format of SQLite. Other applications that use the same format are Calendar, Messages, Notes, and Photos.
- **Recent Items**: References in directory can be found to recently installed or opened applications, servers and files. Recent items in MAC can be typically found in preferences directory: – %%users.homedir%%/Library/Preferences/com.apple.recentitems.plist
- In the same directory, application specific recent exists in the file (LSSharedFileList.plist).
- **Device back-up locations**: Typically, the default location is: %%users.homedir%%/Library/ApplicationSupport/MobileSync/Backup/*
- **System Logs**: There are four main locations for logs: System log files (/var/log/*), Apple system logs (/var/log/asl/*), installation logs (/var/log/install.log) and audit logs (/var/audit/*).
- **Memory SWAP files:** Those are location of virtual memory used to support main physical memory: Those SWAP files can exist in: (/var/vm/sleepimage), (/var/vm/swapfile#).
- **Browser cookies/histories:** Google chrome: %%users.homedir%%/Library/ApplicationSupport/Google\Chrome\Default/*
- **Safari:** %%users.homedir%%/Library/Safari/History.plist/ and LastSession.plist

Cookies can be found in:

%%users.homedir%%/Library/Cookies/Cookies.plist
Cache can be found in: %%users.homedir%%/Library/Caches/com.apple.Safari/
  Cache.db
Visited site at: %%users.homedir%%/Library/*Safari/TopSites.plist*

Skills Section

Disk Forensic Tools

Several forensic activities can be taken when dealing with data in a drive that is inaccessible to the file system. For example "File carving" is a process in which files are reconstructed from raw data (data that has no record or is not extracted based on the file system). Many forensic tools such as: FTK forensic toolkit from accessdata (accessdata.com/solutions/digital-forensics/forensic-toolkit-ftk), EnCase (https://www.guidancesoftware.com/encase-forensic), Foremost (foremost.sourceforge.net), etc. can retrieve raw data based on file carving process. The process accomplishes this file retrieval based on the actual content rather than metadata extracted from file system. Foremost is an open source file carving tool developed by US Air force agents in 1999, Fig. 10.30. It is also included within Kali forensic tools: **http://tools.kali.org/forensics/foremost**. It was also first used to demo file carving in the 2006 forensic challenge: http://www.dfrws.org/2006.

Foremost can be also used to analyze network packets pcap files and extract files directly from this analysis.

Binwalk (http://tools.kali.org/forensics/binwalk, http://binwalk.org/) is another simple open source disk forensic tool. The disk can also extract information from raw images (e.g. bin routers, switches and system firmwares (Fig. 10.31).

```
root@kali:~/Desktop# foremost -h
foremost version 1.5.7 by Jesse Kornblum, Kris Kendall, and Nick Mikus.
$ foremost [-v|-V|-h|-T|-Q|-q|-a|-w-d] [-t <type>] [-s <blocks>] [-k <size>]
          [-b <size>] [-c <file>] [-o <dir>] [-i <file>]

-V  - display copyright information and exit
-t  - specify file type.  (-t jpeg,pdf ...)
-d  - turn on indirect block detection (for UNIX file-systems)
-i  - specify input file (default is stdin)
-a  - Write all headers, perform no error detection (corrupted files)
-w  - Only write the audit file, do not write any detected files to the disk
-o  - set output directory (defaults to output)
-c  - set configuration file to use (defaults to foremost.conf)
-q  - enables quick mode. Search are performed on 512 byte boundaries.
-Q  - enables quiet mode. Suppress output messages.
-v  - verbose mode. Logs all messages to screen
```

Fig. 10.30 Foremost forensic tool

```
root@kali:~# binwalk -B dd-wrt.v24-13064_VINT_mini.bin

DECIMAL       HEX            DESCRIPTION
--------------------------------------------------------------------------------------------------------
0             0x0            TRX firmware header, little endian, header size: 28 bytes, image size: 2945024 bytes, CRC3
28            0x1C           gzip compressed data, from Unix, NULL date: Wed Dec 31 19:00:00 1969, max compression
2472          0x9A8          LZMA compressed data, properties: 0x6E, dictionary size: 2097152 bytes, uncompressed size:
622592        0x98000        Squashfs filesystem, little endian, DD-WRT signature, version 3.0, size: 2320835 bytes,  5
```

Fig. 10.31 Binwalk disk forensic tool

Disk Forensics

Raw Format Image Acquisition

- Conduct a small experiment to show the difference between the 2 tools for image acquisition: dd and dcfldd.
- Complete the steps described in the file [Analysis of hidden data in the NTFS file system] for hiding data in NTFS. Present and submit a report of your activities.

Using Hex-Editors

- Download the picture 10.1. Then used a hex-editor program (e.g. WinHex: http://www.x-ways.net/winhex/) to open and analyze this file. In reality, this file while looks as a simple picture, hides inside another file. Your goal is to extract the second file, find out its type, size (modify its extension to reflect its actual type). Hint (as a jpeg file, check its start and EOF markers to know the end of the first file and copy the rest to a new file. Each file ends with certain hex-markers and then 00000)
- You are supposed now to show the process of embedding an exe file (as an example of possible malware) into an image file. Find proper tools to do this packaging. Then show using a Hex editor tool how you can split the 2 files and extract the exe file from the image file. Document in the report your steps and also attach the streamed file (the image file with an exe embedded into it).

Foremost and File Carving

In this lab, you are expected to search for an image available through the Internet that includes "file carving" issues. Follow the steps similar to the images below to extract raw data files from the image.

- The command below will extract the image to an (out) folder

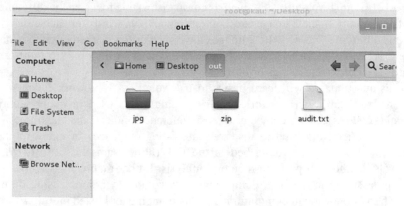

- We can check content of extracted folder. Foremost has predefined folders based on either files extensions (jpg, gif, png, bmp, avi, exe, mpg, wav, riff, wmv, mov, pdf, ole, doc, zip, rar, htm, and cpp) or forensic related information (zip codes, IP addresses, etc.).

- The audit file shows extracted or recovered files from the image

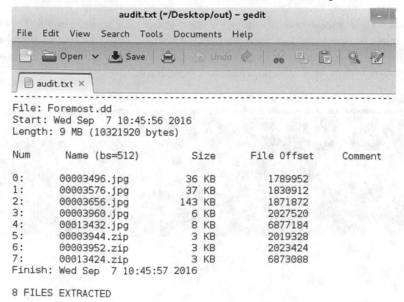

```
File: Foremost.dd
Start: Wed Sep  7 10:45:56 2016
Length: 9 MB (10321920 bytes)

Num      Name (bs=512)           Size        File Offset      Comment

0:       00003496.jpg            36 KB       1789952
1:       00003576.jpg            37 KB       1830912
2:       00003656.jpg            143 KB      1871872
3:       00003960.jpg            6 KB        2027520
4:       00013432.jpg            8 KB        6877184
5:       00003944.zip            3 KB        2019328
6:       00003952.zip            3 KB        2023424
7:       00013424.zip            3 KB        6873088
Finish: Wed Sep  7 10:45:57 2016

8 FILES EXTRACTED
```

- Fine a public pcap file and show how Foremost tool can be used to extract information from this file.

Applications Section

Memory Analysis with Volatility

- Download the images file from https://www.memoryanalysis.net/amf . The folder has 6 Linux Bin sample memories. Pick 2 of those bin Linux memories. Then using Volatility memory analysis tool, make a comparison for each output from the tool (using the exact same commands described in: https://samsclass. info/121/proj/p4-Volatility.htm. In your comparison table focus on the differences in the output from the 2 images explaining in your own words for each case what is this difference and why the 2 images have such difference although they have the exact same operating system. Notice that changes can come from different hardware, software, settings or users.
- In this question, we will learn basics to use Volatility framework for memory forensics. Volatility framework is very popular. It is used in many security training courses. Several computer forensic references indicate that the framework has been effectively used to discover interesting evidences in many forensic cases. Follow the steps described in the link (https://samsclass.info/121/proj/ p4-Volatility.htm) to perform memory analysis. Rather than using the memory dump described in the link, dump your own computer memory using dumpit tool: http://qpdownload.com/dumpit/, or any dumping tool listed in (http://foren-sicswiki.org/wiki/Tools:Memory_Imaging). You can also use publicly available memory dumps (e.g. 1. https://www.memoryanalysis.net/amf) or (2. https:// code.google.com/archive/p/volatility/wikis/SampleMemoryImages.wiki). One example for the second dataset, Stuxnet malware. This is a link describing how Volatility was used to analyse Stuxnet: https://volatility-labs.blogspot. com/2013/05/movp-ii-21-rsa-private-keys-and.html)

```
 C:\Users\ialsmadi\Downloads\DumpIt\DumpIt.exe
DumpIt - v1.3.2.20110401 - One click memory memory dumper
Copyright (c) 2007 - 2011, Matthieu Suiche <http://www.msuiche.net>
Copyright (c) 2010 - 2011, MoonSols <http://www.moonsols.com>


    Address space size:      9099542528 bytes (   8678 Mb)
    Free space size:       122523734016 bytes ( 116847 Mb)

  * Destination = \??\C:\Users\ialsmadi\Downloads\DumpIt\DESKTOP-T2VSKJH-20170430-181012.raw

  --> Are you sure you want to continue? [y/n] y
  + Processing... Success.
```

Linux Logs

Earlier Table, 10.8 shows location of Linux logon log files. In addition to the three mentioned earlier in the Table (i.e. lastlog, messages, wtmp), there are many interesting login log files in (/var./log) directory. Student is encouraged to collect more details about those log files and what information can be extracted from them. They can be all reached using (/var./log) and then one of the following names: messages, dmesg, auth.log, boot.log, daemon.log, dpkg.log, kern.log, lastlog, mail.log, user. log, Xorg.x.log, alternatives, btmp, cups, anaconda.log, yum, cron, secure, wtmp, utmp, failog, httpd, apache2, lighttpd, conman, mail, prelink, audit, setroubleshoot, samba, sa, and sssd. Using your Linux system, show the content of each one of those files.

Kali Linux

One of the popular open source security operating systems is Kali Linux (www.kali. org). You can install Kali in Forensic live mode or as a virtual image. Many Linux forensic tools come pre-installed in Kali: http://docs.kali.org/general-use/kali-linux-forensics-mode. Demo the use of one of the pre-installed forensic tools in Kali. If the tool requires an OS image to use, search for an example of publicly available OS images. Caine is another example of Linux forensic or security operating systems (http://www.caine-live.net/). Other interesting forensic frameworks to investigate include: SIFT (SANS Investigative Forensic Toolkit), DEFT linux (Digital Evidence & Forensics Toolkit), and Martiux.

Windows Registry

Find a Windows XP image online and then search for the following registry keys. For each key, show the results and the implications from forensic perspective:

HKCU\Software\Microsoft\Windows\CurrentVersion\Explorer\ComDlg32\ OpenSaveMRU
HKCU\Software\Microsoft\Windows\CurrentVersion\Explorer\ComDlg32\ LastVisitedMRU
HKCU\Software\Microsoft\Windows\CurrentVersion\Explorer\RecentDocs
HKCU\Software\Microsoft\Windows\CurrentVersion\Explorer\RunMRU
HKLM\SYSTEM\CurrentControlSet\Control\Session   Manager\Memory Management
HKCU \Software\Microsoft\Search Assistant\ACMru
HKLM\SOFTWARE\Microsoft\Windows\CurrentVersion\Uninstall
HKLM \SYSTEM\MountedDevices

HKCU\Software\Microsoft\Windows\CurrentVersion\Explorer\MountPoints2\
CPC\Volume\
HKLM\SYSTEM\CurrentControlSet\Enum\USBSTOR
HKLM\SOFTWARE\Microsoft\Command Processor
HKCU\Software\Microsoft\Command Processor
HKLM\SOFTWARE\Microsoft\Windows NT\CurrentVersion\Winlogon
HKLM\SYSTEM\CurrentControlSet\Services\
HKLM\SOFTWARE\Microsoft\Windows NT\CurrentVersion\Image File Execution
Options\
HKCR\exefile\shell\open\command\
HKLM\SYSTEM\CurrentControlSet\Services\Tcpip\Parameters\Interfaces\GUID
HKLM\SOFTWARE\Microsoft\WZCSVC\Parameters\Interfaces\GUID
HKCU\Software\Microsoft\Windows\CurrentVersion\Explorer\UserAssist
HKCU\Software\Microsoft\Protected Storage System Provider
HKCU\Software\Microsoft\Internet Explorer\TypedURLs

Linux Rootkit Checker

ChkRootkit tests for the presence of certain Rootkits, worms, and Trojans on your system. If you suspect you've been hacked, this is a good first step toward confirmation and diagnosis. Using your Kali image, install Rootkits checker in the example and follow the process to check if you have any Rootkits Malware

- Download the tarfile from http://www.chkRootkit.org, verify its checksum: $ md5sum chkRootkit.tar.gz
- Unpack it: $ tar xvzpf chkRootkit.tar.gz
- Build it: $ cd chkRootkit-*, $ make sense, and run it as root:, #. /chkRootkit

Questions

- Make a comparison between FAT16 and FAT32 showing at least 3 major differences (e.g. additions to FAT32 that were not in FAT16).
- DD is a simple Linux tool to acquire raw image formats. What are the extensions for image raw formats? Describe examples of other tools for raw acquisition.
- What are the major advantages and disadvantages of raw image formats?
- Make a table comparison of strengths and weaknesses between the 3 image acquisition formats: raw, proprietary and independent formats.
- Hashing can be used for integrity checking. Describe major criteria for a hashing algorithm to be effective.
- Can we reverse engineering a hashing value? How is hashing different from encryption?
- Describe 4 examples of major forensic artifacts that can be extracted from any computer case.
- What is slack space? Why it can be relevant to computer forensics?

- Describe how can we use the tool fsstat for disk forensics?
- How can Hex editors be used to detect file types? List examples of 4 popular file types with their start and EOF markers.
- Research for a forensic case in which data hiding was used to cover evidence information. Describe how/what forensic tools were used?
- Why and how much technical detail a forensic investigator needs to know?
- What are the major enhancements of NTFS over FAT?
- Describe some of MFT metafiles and their forensic relevancy
- Can bad sectors be used to hide data? How can such cases be detected?
- What is ADS and how it can be used to hide data?
- Create a report to show how ADS can be created and detected using one of the tools described in the chapter.
- How inodes in ext. FS are compared to clusters in FAT or NTFS?
- What are the advantages and disadvantages of enabling the journaling process?
- Describe how you can use mmls tool in a forensic analysis.
- Why it's harder to conduct memory forensic in comparison with disk forensics
- Describe 4 examples of forensic relevant information that can be found in memory.
- Show using s demo report how can you dump a disk memory image to a file and then extract/analyze extracted memory file.
- Show in basic steps how you can use "volatility" memory analysis tool to analyze a memory image.
- Use an example of Windows image available online to show all artifacts described in Table 10.6.
- Does Linux have registry database like Windows? If not, how we can extract similar registry information from Linux?
- What Internet traces can be analyzed from a suspect disk? If users cleaned their history, can we still detect such information?
- What are "cookies" and how they are relevant for forensic investigations?
- Describe examples where event viewers' information can be relevant to forensic investigations.
- What information about a client machine can be found in a web log? How can such information be correlated with user local Internet history?
- Use your Kali Linux image to show content of directories describe in Table 10.7.
- Use your Kali Linux image to show content of directories describe in Table 10.8.
- Use a MAC disk image (from public resources) to show main forensic artifacts described in the chapter.

References

https://www.memoryanalysis.net/amf
http://www.windowsscope.com
https://samsclass.info/121/proj/p4-Volatility.htm

http://www.forensicswiki.org/wiki/Linux_Memory_Analysis
http://www.volatilityfoundation.org/
https://www.guidancesoftware.com/encase-forensic
http://www.dfrws.org/2006
http://www.x-ways.net/winhex/
https://code.google.com/p/volatilitux
https://www.nsrl.nist.gov/Downloads.htm
accessdata.com/solutions/digital-forensics/forensic-toolkit-ftk
https://www.guidancesoftware.com/encase-forensic
http://www.linuxleo.com/
www.kali.org
https://forensics.cert.org/
https://github.com/sans-dfir/sift
http://www.caine-live.net/
SANS. (2011). Digital Forensic SIFTing - Mounting Evidence Image Files, 28-Nov-2011,
    https://digitalforensics.sans.org/blog/2011/11/28/digital-forensic-sifting-mounting-ewf-or-
    e01-evidence-image-files/commentpage-1/

Chapter 11
Network Forensics: Lesson Plans

Competency: Learn major aspects of switches' forensics
Activities/Indicators
• Study reading material provided by instructor related to major aspects of switches' forensics • Complete successfully an assessment provided by instructor related to competency content. • For mastering levels, more than 80% of assessment grades should be earned in no more than three trials. • Assessment questions can be pulled from the end of the chapter questions or any relevant material.
Competency: Learn major aspects of routers' forensics
Activities/Indicators
• Study reading material provided by instructor related to major aspects of routers' forensics • Complete successfully an assessment provided by instructor related to competency content. • For mastering levels, more than 80% of assessment grades should be earned in no more than three trials. • Assessment questions can be pulled from the end of the chapter questions or any relevant material.
Competency: Learn major aspects of IDS/IPS's forensics
Activities/Indicators
• Study reading material provided by instructor related to major aspects of IDS/IPS's forensics • Complete successfully an assessment provided by instructor related to competency content. • For mastering levels, more than 80% of assessment grades should be earned in no more than three trials. • Assessment questions can be pulled from the end of the chapter questions or any relevant material.
Competency: Learn major aspects of wireless' forensics
Activities/Indicators
• Study reading material provided by instructor related to major aspects of wireless' forensics • Complete successfully an assessment provided by instructor related to competency content. • For mastering levels, more than 80% of assessment grades should be earned in no more than three trials. • Assessment questions can be pulled from the end of the chapter questions or any relevant material.

(continued)

© Springer International Publishing AG 2018
I. Alsmadi et al., *Practical Information Security*,
https://doi.org/10.1007/978-3-319-72119-4_11

Competency: Learn how to conduct switches' forensics.

Activities/Indicators

- Use one or more open source or free tools that allow users to evaluate switches' forensics.
- For mastery, student is expected to try more than one switch type.
- For a mastering level, student should show what tool they selected, how they installed the tool and/or present a video to demonstrate the different commands they have used.

Competency: Learn how to conduct routers' forensics.

Activities/Indicators

- Use one or more open source or free tools that allow users to evaluate routers' forensics.
- For mastery, student is expected to try more than one switch type.
- For a mastering level, student should show what tool they selected, how they installed the tool and/or present a video to demonstrate the different commands they have used.

Competency: Learn how to conduct IDS/IPSs' forensics.

Activities/Indicators

- Use one or more open source or free tools that allow users to evaluate IDS/IPSs' forensics.
- For mastery, student is expected to try more than one switch type.
- For a mastering level, student should show what tool they selected, how they installed the tool and/or present a video to demonstrate the different commands they have used.

Competency: Learn how to conduct wireless' forensics.

Activities/Indicators

- Use one or more open source or free tools that allow users to evaluate wireless' forensics.
- For mastery, student is expected to try more than one switch type.
- For a mastering level, student should show what tool they selected, how they installed the tool and/or present a video to demonstrate the different commands they have used.

Competency: Learn how to extract forensics artifacts from a network components

Activities/Indicators

- Conduct a case on how to how to extract forensics artifacts from a network components.
- For a mastering level, student is expected to show more than one OBAC example or application.
- For a mastering level, student should show what tool they selected, how they installed the tool and present a video to demonstrate the different commands they have used.

Overview

A forensic investigator who is analyzing computer equipment for possible evidences, will search different locations for possible traces. We described in other chapters the types of evidences that can be found in disks or operating systems. There are some network or Internet traces that can be found in Internet browsers' history. From an OSI perspective, such information is typically in the higher layers (i.e. layer 7). Network forensics focus on searching, monitoring and/or analyzing network components, (i.e. switches, routers, firewalls, wireless, Intrusion detection/prevention systems IDS/IPS) for possible forensic evidences. In many cases, it is important to correlate some information from a host with information collected from the network to make sure that a host or some of its artifacts were not tampered by suspect or intruders.

We will divide this chapter based on those five previously mentioned components.

Knowledge Section

Traffic Analysis

Regardless of the source that logged the network traffic, there is a general standard of what traffic attributes can be collected. Traffic includes flows of data/information from a source to a destination. The details of information depend on the OSI level through which information is investigation. We will have a quick look on OSI 7 layers. Students are requested to read more details on OSI model through different resources or references. Figure 11.1 shows OSI 7 layers. The first layer from the users' side is the application layer. The first layer from the network/medium side is the physical layer.

It is important for network forensic analyst to understand those different layers, the nature of data in each layer and that type of protocols or applications that are relevant to each layer. As an alternative model, TCP/IP model (Fig. 11.2) has 4 layers in which the 3 OSI layers: Application, presentation and session are combined in one layer: application layer. Figure 11.2 shows also some of the popular protocols and their location in the two models.

In comparison with disk and OS forensics, the number of protocols and applications in network forensics is much larger. Additionally, information can be always mixed in network forensics where many protocols and applications can be seen in one network log or traffic. Additionally, those applications and protocols change frequently much more than file or operating systems. Protocols or standards are not enforced in the network and hence different implementations from different companies can be different for the same networking protocol.

Network forensics involves steps to search through the public Internet. Unlike desk or OS forensics, network forensics territory is large, open and non-centralized.

Fig. 11.1 OSI Layers

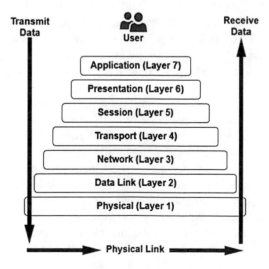

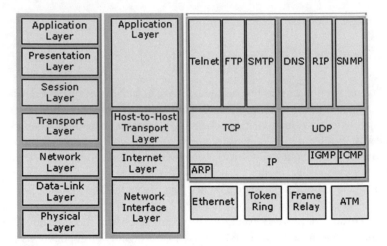

Fig. 11.2   TCP/IP model

Investigators may need to use different types of tools/mechanisms in the same investigation. They may need to use techniques that are "outside the box". They may need to communicate with local, national or international websites or organizations. Legal and technical barriers and consideration can be very serious.

One more challenge in network forensic is the mixture of IPv4 and IPv6 traffic. While IPv6 starts a couple of decades ago, yet more than three quarters of Internet traffic is IPv4. Any instance of traffic analysis can have a mixture of both. Same applications or protocols can have different views in those two IP addressing schemes. Network forensic analysts should also familiarize themselves with those two IP addressing schemes, the differences between them and how to distinguish attributes/protocols related to each one of them. The important of IP addresses is that they are the single information from all traffic information that can uniquely identify the source and the destination of the traffic. Once we know the IP address (i.e. the desktop, laptop, smart phones, etc.) that was the source or the destination of the traffic, we can eventually know the user. Can IP addresses be faked or changed? Yes. IP spoofing is popular and its widely used for different types of purposes or goals (i.e. hide identity, access resources that cannot be accessed using actual IP addresses, etc.). We can also use network analysis to investigate if there are indicators of the occurrence of IP, or MAC spoofing.

There are some similarities between network and memory forensics. Both are transient; do not employ persistent or static storage. Additionally, for performance reasons, memory and network data has limited space to store history. Similar to seizing a memory, seizing a network evidence can be also troublesome. It may require pulling a device that is heavily used. As a general guideline in forensics, it is recommended to collect evidences, as soon as possible. Hashing is used in desk forensics to preserve the integrity of the evidence and make sure it is not tampered after seizure. Can we implement hashing for network types of evidences.

From where to start a network forensic investigation? We can divide forensic investigations from this perspective into two categories:

- **Top Down**; where we have questions and want to use network forensics in trying to answer those questions. Let's say for example that we have a server intrusion problem. Somebody hacked yesterday into our server. So we started by raising some question like: How do we know an intrusion happened? When did the intrusion happen? Who was the source of the intrusion? What they did/changed/deleted/etc. on our server? What vulnerability or weaknesses caused that intrusion? How can we prevent similar intrusions in the future? Ultimately, we want to use all forensic methods (i.e. disk, OS, network, etc. forensics) to be able to answer all those questions.
- **Bottom up**: We have some systems, servers, or network monitoring tools that showed us a possible intrusion or some abnormal behaviors. As a result, we started our forensic investigation looking through the output of those monitoring tools (i.e. network traffic or logs) in order to see what happened. We may raise the same earlier questions. In the first type, we may also end up using the same outputs of those monitoring tools. The goals are the same, but from where we start or some activities can be different.

Traffic can be analyzed either in real time (i.e. from the network media: cables, air, etc.) or analyzed statically based on collected traffic history from network components.

Real Time Network Forensic Monitoring or Analysis

If suspicious events are spotted in a network, it may be appropriate to conduct real time analysis of network traffic. This can be significantly different from classical network traffic monitoring in which all traffic details are logged. It can be also different from the work of a firewall or an IDS/IPS in which roles are built to control: permit/deny traffic in real time based on different types of policies. Network forensic monitoring will be much more focus in comparison with those real time monitoring or traffic control mechanisms/tools. For example, if there is a possible malware or intrusion/hacking scheme that is suspected in a network, special monitoring tools will be interested on looking at certain type of traffic: coming from certain sources, targeting certain destinations, using certain ports or any special known attributes that are known so far about this attack scheme.

Static Network Forensic Analysis

Typically, this happens when we have an incident in the past and we are conducting a network forensic analysis to collect some evidences and know what happened. In other words, we know the date, possibly the approximate time, or period of the incident. We have a dataset of web logs or network traffic that we want to use it as the input to the investigation process.

DHCP Servers

Many network functionalities can be defined in regular computers or servers (i.e. not in dedicated network components: switches, routers, etc.). DHCP servers are good examples of such network servers typically deployed on computing servers. DHCP servers assign private IP addresses dynamically to local hosts. If we have an IP address in a forensic case within a certain day or period, we need to refer to the DHCP server log to know who (i.e. which machine, MAC or physical address) was given this "suspect" IP addresses in the "suspect" date and time.

Similar to DHCP servers, there are many different network services that may leave forensic traces such as: Access, authentication or identity servers, name or DNS servers, email servers, web servers, database servers, etc.

Wire Tapping

For a telephone network, wire tapping is like network sniffing which involves listening to the incoming/outgoing data in the telecommunication wire or media. While, literally, cutting a wire to listen to it (i.e. vampire or inline tapping) can be rare and unconventional, yet there are many methods that allow "virtual tapping". For example, induction coils can be used to tap into coaxial cables. This can be much harder if cable is shielded. Tapping tools can be also located in connection points: inline-tapping (e.g. switches, routers, etc.). With fiber optic, inline tapping is impractical and may cause significant signal degradation.

The most currently used wired telecommunication media or cables include: Fiber optic, twisted pair and coaxial cables.

Wireless Tapping

Wireless tapping is more popular and easier to implement. Additionally, it will be hard to detect, especially if it is passive tapping (i.e. without sending any sort of signal, traffic or data).

Network sniffing can be part of normal network activities (e.g. for security, maintenance, access control, troubleshooting). They don't have to be triggered occasionally or once an incident occurs. Nonetheless, the nature and details of what a specific forensic case wants to sniff can be different from the "normal" sniffing activities. Extra or more details on certain hosts, targets, applications, etc. can be requested.

Spying in WiFi networks is very popular. For home users and their wireless access points, there are different types of challenges or problems:

- The usage of old or unreliable encryption algorithms or schemes. For example, in most home wireless routers, it is recommended to use (Wireless Access Point)

WAP or WAP2 as alternatives to the old (Wired Equivalent Privacy) WEP encryption method which seems to be much easier to crack or decrypt passwords from. Most home wireless routers or access points still support WEP and may use this encryption method as the default.

- Generic or simple SSIDs and/or passwords. Hackers or users who search a neighborhood for available WiFi or wireless networks will see the broadcasted name of the WiFi network or the: SSID. One problem is that each manufacturing company has certain naming standards. Users who choose to use those default naming standards can make it easy for hackers to crack their networks. For one point, hackers can then easily know the specific manufacture of the access point and some other important relevant information (e.g. encryption methods). They can use this as a starting point to focus their attacking effort. Additionally, those names can be reused. For example, if your smart phone once verified the credentials of a public WiFi network, a hacker can create a trap WiFi with the same name or SSID. Your phone will automatically recognize and connect to such network. The hacker can then use some tools to monitor and spy on your traffic, steal your credentials, etc.
- WiFi access points can be conFigured with weak, trivial or dictionary-based passwords. Cracking tools can be used to crack or guess such passwords and access, or at least, use this WiFi network. In the current network architecture, it will be very hard to, forensically, differentiate who was using the WiFi; the legitimate user or the hacker.

There are some tools and methods not only to detect broadcasted SSIDs or other WiFi information, but also un-broadcasted ones.

Ethernet Cards Promiscuous Mode

Promiscous or promisc mode is used in Ethernet Network cards (NIC) to allow monitoring network traffic. In this mode, the NIC will pass all the traffic it receives to a central controller, rather than just passing the traffic meant to be for the NIC or interface. In normal, non-promisc mode, this NIC or interface will drop all traffic not intended for it or if it is a broadcasting message (i.e. a message to all interfaces). This NIC mode (i.e. promisc mode) can be used for network packet sniffing, monitoring, etc.

Promisc mode can be used as part of a malicious activity or attack. In some forensic cases, it is very important to check of any NIC in the network is running in a promisc mode. In wireless network, where users can register and register quickly and they don't have cables to connect with the network, detection of such behavior can be hard. A port in a hub by default is in "promisc mode" and will transmit received traffic to other ports. Having a hub in a network can give an advantage to network forensic investigators as they can listen to network without much effort. However, they need to be careful as their traffic can be also seen by others.

Wireshark

One of the most important skills in network security is the ability to extract knowledge from network traffic. Network traffic can be intercepted at the network level from network devices, but can also be intercepted at the host level for incoming and outgoing traffic for a particular host or computer. In many forensic cases, we have one suspect machine to monitor and hence we can use host-based network traffic. For verification and integrity, it can be also correlated with network traffic.

Wireshark, originally Ethereal, is a popular tool used for traffic analysis (Fig. 11.3). Other tools with similar functions include: tcpdump, dumPCAP, windump, tshark, iftop, P0f, Snort, nmap, ngrep, etc. Due to the significant similarity between those different tools, we will use Wireshark to demo this section. Student is encouraged to experience the other tools as well.

Input files that include traffic information can have extensions like .PCAP, .PCAPng, etc.

LibPCAP and WinPCAP are libraries in Linux and Windows for traffic capturing.

Each record in those files represents a packet. Packets are in a time sequence (relative time based on when recording was started). For each packet, investigator can see different details based on the OSI layers (i.e. Frame, Ethernet, logical or layer 3, etc.).

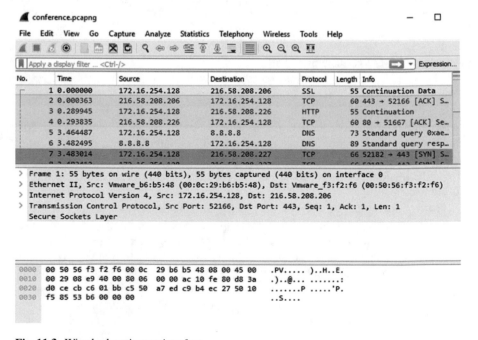

Fig. 11.3 Wireshark main user interface

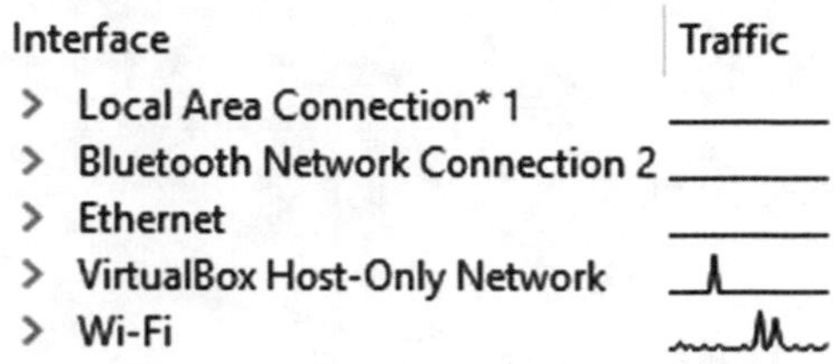

Fig. 11.4 Wireshark
possible sniffing interfaces

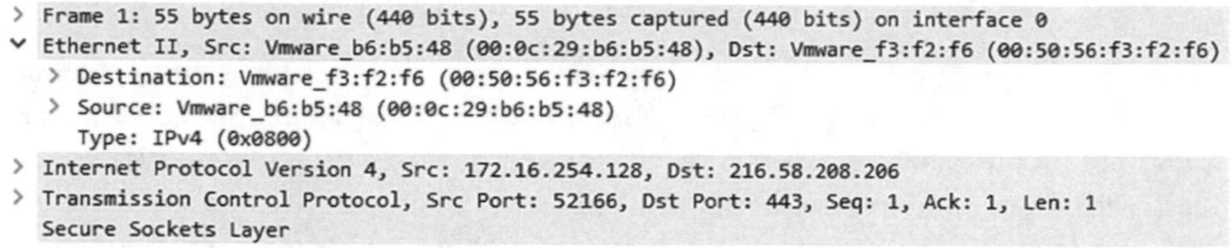

Interface	Traffic
> Local Area Connection* 1	____
> Bluetooth Network Connection 2	____
> Ethernet	____
> VirtualBox Host-Only Network	____
> Wi-Fi	____

```
> Frame 1: 55 bytes on wire (440 bits), 55 bytes captured (440 bits) on interface 0
v Ethernet II, Src: Vmware_b6:b5:48 (00:0c:29:b6:b5:48), Dst: Vmware_f3:f2:f6 (00:50:56:f3:f2:f6)
  > Destination: Vmware_f3:f2:f6 (00:50:56:f3:f2:f6)
  > Source: Vmware_b6:b5:48 (00:0c:29:b6:b5:48)
    Type: IPv4 (0x0800)
> Internet Protocol Version 4, Src: 172.16.254.128, Dst: 216.58.208.206
> Transmission Control Protocol, Src Port: 52166, Dst Port: 443, Seq: 1, Ack: 1, Len: 1
  Secure Sockets Layer
```

Fig. 11.5 An example of a packet details

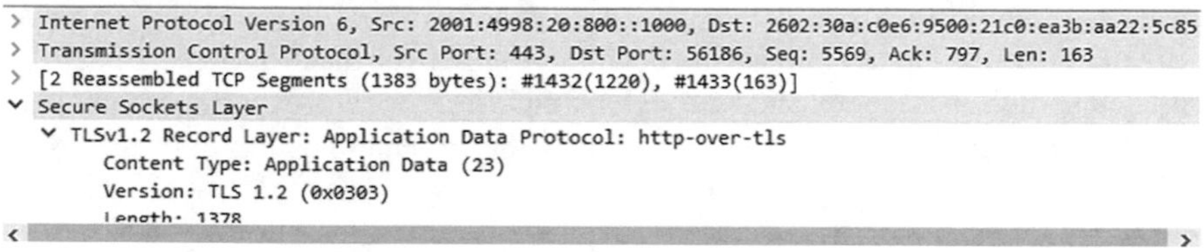

```
> Internet Protocol Version 6, Src: 2001:4998:20:800::1000, Dst: 2602:30a:c0e6:9500:21c0:ea3b:aa22:5c85
> Transmission Control Protocol, Src Port: 443, Dst Port: 56186, Seq: 5569, Ack: 797, Len: 163
> [2 Reassembled TCP Segments (1383 bytes): #1432(1220), #1433(163)]
v Secure Sockets Layer
  v TLSv1.2 Record Layer: Application Data Protocol: http-over-tls
      Content Type: Application Data (23)
      Version: TLS 1.2 (0x0303)
      Length: 1378
<
                                                                                              >
```

Fig. 11.6 SSL encrypted packets

You can use Wireshark to read PCAP files recorded previously. It can be also used to record traffic in real time from any type of interface, Fig. 11.4 (Wired, wireless, Bluetooth, etc.).

Analyzing traffic for possible evidences may require different types of technical skills. Analyst should know about the different protocols that exist in the network traffic. They need to know how they behave, normally. They need to know "abnormal" behaviors. They can look up details on network protocols from many references (e.g. internet Engineering Task Force: IETF: https://www.ietf.org/ , RFCs, IEEE, WWW, ISO, etc.). One challenge in this regard is that different companies may not follow the exact algorithm specifications. Such, vendor related variations should be also known to forensic analysts. In some cases, they may not find such variations published publicly. Many network analyst tools such as Wireshark provide protocol dissectors for many protocols which can be very help in protocols' identification. Users can also add/define new protocols' dissectors. Figure 11.5 shows details on one packet. From Ethernet II section, we can see that this is an IPv4 packet. We can see also the source and destination host names and MAC addresses. From the TCP (or UDP) section we can see this type of protocol and also source and destination port numbers. In many cases, knowing source and destination port numbers can help identifying the application. The packet below for example shows destination port 443. If we lookup this port number we can see that it is a typical indicator for encrypted (HTTPS) traffic. Port 53 is known for DNS queries (Fig. 11.6).

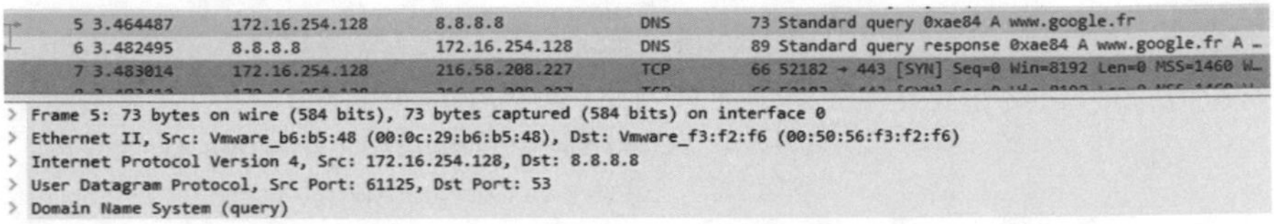

```
   5 3.464487      172.16.254.128      8.8.8.8           DNS      73 Standard query 0xae84 A www.google.fr
   6 3.482495      8.8.8.8             172.16.254.128    DNS      89 Standard query response 0xae84 A www.google.fr A -
   7 3.483014      172.16.254.128      216.58.208.227    TCP      66 52182 → 443 [SYN] Seq=0 Win=8192 Len=0 MSS=1460 W_

> Frame 5: 73 bytes on wire (584 bits), 73 bytes captured (584 bits) on interface 0
> Ethernet II, Src: Vmware_b6:b5:48 (00:0c:29:b6:b5:48), Dst: Vmware_f3:f2:f6 (00:50:56:f3:f2:f6)
> Internet Protocol Version 4, Src: 172.16.254.128, Dst: 8.8.8.8
> User Datagram Protocol, Src Port: 61125, Dst Port: 53
> Domain Name System (query)
```

Fig. 11.7 A DNS query using port 53

Using HTTPS protocol by itself may not ensure that traffic is encrypted or secure. In Fig. 12.5, we see that there is a menu below SSL (Secure Sockets Layer), the encryption protocol; however, it has no content. If this is properly encrypted, we should see content under this SSL section (Fig. 11.7).

We can learn also from Fig. 11.5 Ethernet part that both source and destination hosts are virtual machines running VMware operating systems.

In many cases network analyst may need to recreate a scenario of activities from the user through looking at the sequence of packets rather than looking at each individual packet. Figure 11.8 shows a sample output for the sequence of packets with details. This is accomplished through using Wireshark and exporting the PCAP file into PDML XML output. Users can specify the amount of details to be included in the exported file.

Packet Filtering

The forensic search through a large PCAP file with many packets can be very tedious and time consuming. One of the methods to perform quick or short searches is to use filters. Wireshark includes as a rich menu for filtering expressions based on a large set of network related attributes (Fig. 11.9).

Wireshark filters allow analysts to search using protocols with expression to match certain criteria or follow. Analysts can also combine more than one criterion in their filter (Fig. 11.10).

Possibly I am investigating a malware and I am focusing on looking for large packets or frame. I can use a filter such as: frame.cap_len >= 2000 (or any relevant size); Figs. 11.11 and 11.12.

Using the "right" filters can be context-dependent. Forensic analyst should first read about the case and what we are looking for or trying to find. They may start by writing some questions and then trying to answer them based on investigation packets using different filters. For example, if we have a question such as: What HTTP packets suspect host sent? We may use the filters below to show the output: tcp.port == 80 && ip.src == 192.168.40.10 (Fig. 11.13). You may use the feature: Follow stream which will perform something similar (Fig. 11.14). Follow TCP stream can be very useful for forensic analyst to help them see a complete readable stream rather than digging into packets, which can be very time consuming.

```
<?xml version="1.0"?>
<psml version="0" creator="wireshark/2.2.6">
<structure>
<section>No.</section>
<section>Time</section>
<section>Source</section>
<section>Destination</section>
<section>Protocol</section>
<section>Length</section>
<section>Info</section>
</structure>

<packet>
<section>1</section>
<section>0.000000</section>
<section>2wire_2a:6b:89</section>
<section>Broadcast</section>
<section>ARP</section>
<section>60</section>
<section>Who has 192.168.1.80? Tell 192.168.1.254</section>
</packet>

<packet>
<section>2</section>
<section>0.102891</section>
<section>192.168.1.66</section>
<section>239.255.255.250</section>
<section>UDP</section>
<section>690</section>
<section>1063 \xe2\x86\x92 8082 Len=648</section>
</packet>

<packet>
<section>3</section>
<section>0.125726</section>
<section>2607:f8b0:4003:c07::bd</section>
<section>2602:30a:c0e6:9500:21c0:ea3b:aa22:5c85</section>
<section>QUIC</section>
<section>102</section>
<section>Payload (Encrypted), PKN: 70</section>
```

Fig. 11.8 A sample PDML XML output from a PCAP file

From Fig. 11.14, we can see for example that TCP stream number 1 has 19 client packets and 19 server packets. The number of packets can vary from one stream to another.

Name resolution tools in Wireshark can help us associate one network information with another. For example, we can associate a MAC address with an IP address or a host or DNS name. It can be also associating a port number to an application. You can enable name resolutions in Wireshark from Capture menu options; Fig. 11.15. The name resolution is not always correct and users should verify it manually.

Forensic analyst may want also to check traffic statistics as they may show some useful information relevant to the investigation. This can be for example certain large packets, or large number of certain packet types (e.g. DNS, ARP, etc.); Fig. 11.16, or some statistics about end-points; Fig. 11.17.

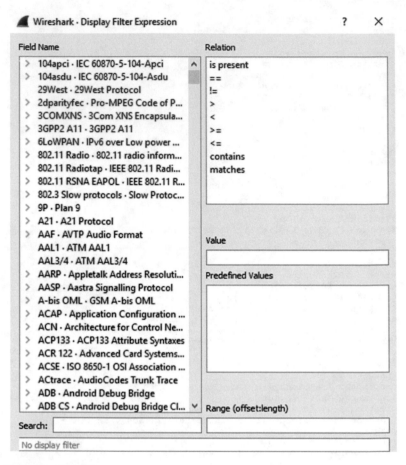

Fig. 11.9 Wireshark filters

No.	Time	Source	Destination	Protocol	Length	Info
2	0.000416	192.168.40.1	192.168.40.10	DHCP	342	DHCP Offer - Transacti...
4	0.001211	192.168.40.1	192.168.40.10	DHCP	342	DHCP ACK - Transacti...
5	2.563151	192.168.40.10	192.168.40.1	DHCP	356	DHCP Request - Transacti...
6	2.563579	192.168.40.1	192.168.40.10	DHCP	342	DHCP ACK - Transacti...

File   Edit   View   Go   Capture   Analyze   Statistics   Telephony   Wireless   Tools   Help

ip.addr == 192.168.40.1

Fig. 11.10 A simple example of Wireshark filters

For example, when we observer endpoints in Fig. 11.17, we can see that this stream is largely a communication between two MACs or hosts. The few packets of other MACs may represent things like a DNS server, etc. Maybe quickly viewing end-points can help us narrow down the problem or the focus on few packets to analyze thoroughly.

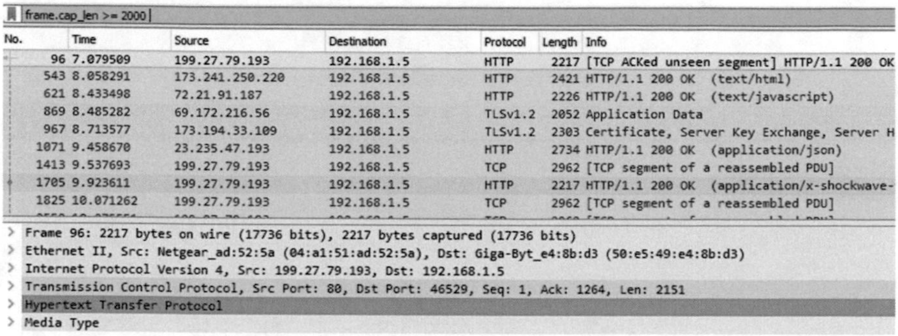

frame.cap_len >= 2000

No.	Time	Source	Destination	Protocol	Length	Info
96	7.079509	199.27.79.193	192.168.1.5	HTTP	2217	[TCP ACKed unseen segment] HTTP/1.1 200 OK
543	8.058291	173.241.250.220	192.168.1.5	HTTP	2421	HTTP/1.1 200 OK (text/html)
621	8.433498	72.21.91.187	192.168.1.5	HTTP	2226	HTTP/1.1 200 OK (text/javascript)
869	8.485282	69.172.216.56	192.168.1.5	TLSv1.2	2052	Application Data
967	8.713577	173.194.33.109	192.168.1.5	TLSv1.2	2303	Certificate, Server Key Exchange, Server H
1071	9.458670	23.235.47.193	192.168.1.5	HTTP	2734	HTTP/1.1 200 OK (application/json)
1413	9.537693	199.27.79.193	192.168.1.5	TCP	2962	[TCP segment of a reassembled PDU]
1705	9.923611	199.27.79.193	192.168.1.5	HTTP	2217	HTTP/1.1 200 OK (application/x-shockwave-
1825	10.071262	199.27.79.193	192.168.1.5	TCP	2962	[TCP segment of a reassembled PDU]

> Frame 96: 2217 bytes on wire (17736 bits), 2217 bytes captured (17736 bits)
> Ethernet II, Src: Netgear_ad:52:5a (04:a1:51:ad:52:5a), Dst: Giga-Byt_e4:8b:d3 (50:e5:49:e4:8b:d3)
> Internet Protocol Version 4, Src: 199.27.79.193, Dst: 192.168.1.5
> Transmission Control Protocol, Src Port: 80, Dst Port: 46529, Seq: 1, Ack: 1264, Len: 2151
> Hypertext Transfer Protocol
> Media Type

Fig. 11.11 A simple example Wireshark filters (filter for large packets)

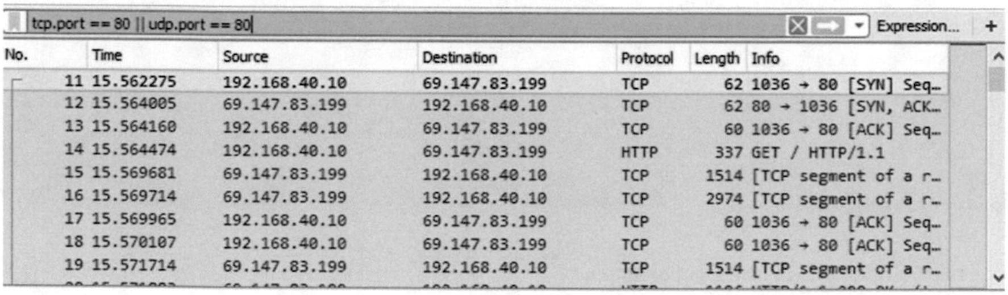

tcp.port == 80 || udp.port == 80 Expression... +

No.	Time	Source	Destination	Protocol	Length	Info
11	15.562275	192.168.40.10	69.147.83.199	TCP	62	1036 → 80 [SYN] Seq…
12	15.564005	69.147.83.199	192.168.40.10	TCP	62	80 → 1036 [SYN, ACK…
13	15.564160	192.168.40.10	69.147.83.199	TCP	60	1036 → 80 [ACK] Seq…
14	15.564474	192.168.40.10	69.147.83.199	HTTP	337	GET / HTTP/1.1
15	15.569681	69.147.83.199	192.168.40.10	TCP	1514	[TCP segment of a r…
16	15.569714	69.147.83.199	192.168.40.10	TCP	2974	[TCP segment of a r…
17	15.569965	192.168.40.10	69.147.83.199	TCP	60	1036 → 80 [ACK] Seq…
18	15.570107	192.168.40.10	69.147.83.199	TCP	60	1036 → 80 [ACK] Seq…
19	15.571714	69.147.83.199	192.168.40.10	TCP	1514	[TCP segment of a r…

Fig. 11.12 A simple example Wireshark filters (more than one criterion)

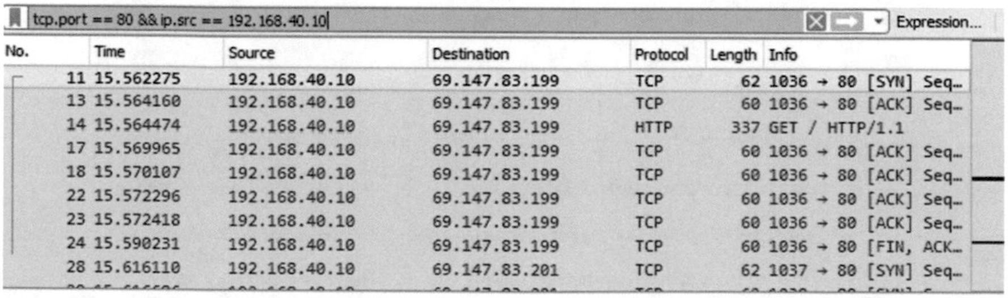

tcp.port == 80 && ip.src == 192.168.40.10 Expression...

No.	Time	Source	Destination	Protocol	Length	Info
11	15.562275	192.168.40.10	69.147.83.199	TCP	62	1036 → 80 [SYN] Seq…
13	15.564160	192.168.40.10	69.147.83.199	TCP	60	1036 → 80 [ACK] Seq…
14	15.564474	192.168.40.10	69.147.83.199	HTTP	337	GET / HTTP/1.1
17	15.569965	192.168.40.10	69.147.83.199	TCP	60	1036 → 80 [ACK] Seq…
18	15.570107	192.168.40.10	69.147.83.199	TCP	60	1036 → 80 [ACK] Seq…
22	15.572296	192.168.40.10	69.147.83.199	TCP	60	1036 → 80 [ACK] Seq…
23	15.572418	192.168.40.10	69.147.83.199	TCP	60	1036 → 80 [ACK] Seq…
24	15.590231	192.168.40.10	69.147.83.199	TCP	60	1036 → 80 [FIN, ACK…
28	15.616110	192.168.40.10	69.147.83.201	TCP	62	1037 → 80 [SYN] Seq…

Fig. 11.13 An example of combined Wireshark filters

For example, Fig. 11.18 shows end-points statistics for another stream. Maybe our focus is on the end-point which has only one packet in all this large stream, why only one packet? , what was the purpose of this single packet? (Fig. 11.19).

Conversation from statistics can serve the same goal of end-points to summarize the main communication sessions in the stream; Fig. 11.20. If we check

If we check or select the (name resolution) option, this can help us better understand communication from hosts or names perspective; Fig. 11.20.

You can specify more details on those statistics using options. For example, you can specify, other types of conversations (e.g. Bluetooth, IEEE 802.11, etc.). Figure 11.21 shows an example of hotspot or WiFi conversation. We can see many

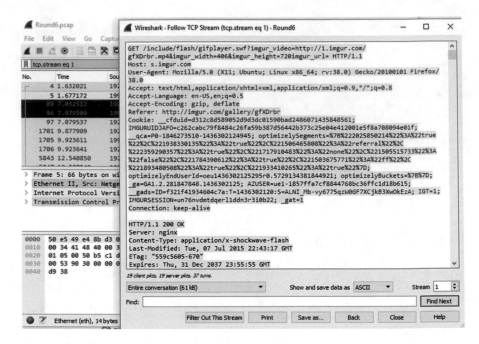

Fig. 11.14 Follow a TCP stream example

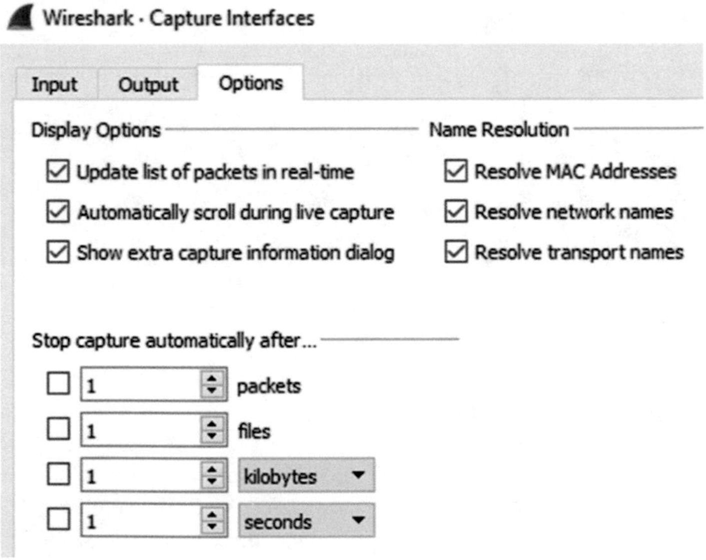

Fig. 11.15 Enable Name resolutions

Protocol	Percent Packets	Packets	Percent Bytes	Bytes	Bits/s	End Packets	End Bytes	End Bits/s
˅ Frame	100.0	15	100.0	3374	2708	0	0	0
˅ Ethernet	100.0	15	6.2	210	168	0	0	0
˅ Internet Protocol Version 4	100.0	15	8.9	300	240	0	0	0
˅ Transmission Control Protocol	100.0	15	83.8	2828	2270	11	665	533
˅ Hypertext Transfer Protocol	26.7	4	74.5	2512	2016	2	605	485
Portable Network Graphics	6.7	1	6.1	207	166	1	422	338
Line-based text data	6.7	1	37.7	1271	1020	1	1485	1192

Fig. 11.16 Packets' statistics in Wireshark

Ethernet · 5	IPv4 · 74	IPv6	TCP · 40	UDP · 79		
Address	Packets	Bytes	Tx Packets	Tx Bytes	Rx Packets	Rx Bytes
00:20:18:eb:ca:28	1,403	1234 k	830	58 k	573	
01:00:5e:00:00:16	2	120	0	0	2	
01:00:5e:7f:ff:fa	3	525	0	0	3	
0a:b4:df:27:c2:b0	1,396	1232 k	573	1175 k	823	
ff:ff:ff:ff:ff:ff	2	710	0	0	2	

Fig. 11.17 Packets' statistics in Wireshark

Ethernet · 8	IPv4 · 277	IPv6 · 2	TCP · 1986	UDP · 845		
Address	Packets	Bytes	Tx Packets	Tx Bytes	Rx Packets	Rx Bytes
00:0b:cd:c2:e4:91	64	5639	50	4571	14	
01:00:5e:00:00:01	29	1740	0	0	29	
01:00:5e:00:00:16	90	5580	0	0	90	
04:a1:51:ad:52:5a	161,648	130 M	92,937	121 M	68,711	
33:33:00:00:00:01	1	102	0	0	1	
50:e5:49:e4:8b:d3	170,538	139 M	71,847	9525 k	98,691	
f0:76:1c:31:be:25	8,944	8983 k	5,828	8774 k	3,116	
ff:ff:ff:ff:ff:ff	10	1278	0	0	10	

Fig. 11.18 Packets' statistics; end-points in Wireshark

single-packet conversations in the Ethernet-tab. If we look at the IP-tab we will see a different view.

If we suspect a port scanning activity, conversation-TCP tab can be a useful indicator. In Fig. 11.22, we can see that while source port is not changing, destination port is changing per sequence packets.

Foremost

Foremost is an open source tool developed original for disk forensics (http://foremost.sourceforge.net/). One of the main distinguished features of Foremost was the ability to analyze raw disk images (Fig. 11.23). In particular the tool is able to handle file or data carving; extract data from unallocated disks or disk partitions.

Ethernet · 13	IEEE 802.11	IPv4 · 281		IPv6 · 1	TCP · 1708	UDP · 841					
Address A	Address B	Packets	Bytes	Packets A→B	Bytes A→B	Packets B→A	Bytes B→A	Rel Start	Duration	Bits/s A→B	Bits/s B→A
HewlettP_c2:e4:91	IPv4mcast_16	30	1860	30	1860	0	0	1.387136	727.0024	20	
HewlettP_c2:e4:91	Broadcast	6	749	6	749	0	0	6.547362	466.1334	12	
HewlettP_c2:e4:91	Netgear_ad:52:5a	24	2738	12	1754	12	984	6.547499	340.8939	41	
HewlettP_c2:e4:91	Giga-Byt_e4:8b:d3	4	292	2	208	2	84	163.906844	308.7740	5	
IPv4mcast_01	Netgear_ad:52:5a	29	1740	0	0	29	1740	21.044454	700.3667	0	
IPv4mcast_16	Netgear_ad:52:5a	30	2100	0	0	30	2100	1.996130	723.6826	0	
IPv4mcast_16	Giga-Byt_e4:8b:d3	30	1620	0	0	30	1620	5.024018	725.6320	0	
Netgear_ad:52:5a	Giga-Byt_e4:8b:d3	161,560	130 M	92,863	121 M	68,697	9315 k	0.000000	731.4960	1327 k	
Netgear_ad:52:5a	Compalln_31:be:25	3	228	1	84	2	144	328.443071	0.0028	—	
Netgear_ad:52:5a	Broadcast	1	60	1	60	0	0	328.445624	0.0000	—	
Netgear_ad:52:5a	IPv6mcast_01	1	102	1	102	0	0	504.286065	0.0000	—	
Giga-Byt_e4:8b:d3	Compalln_31:be:25	8,941	8983 k	3,115	208 k	5,826	8774 k	18.317878	631.9327	2639	
Giga-Byt_e4:8b:d3	Broadcast	3	469	3	469	0	0	163.906639	480.9379	7	

Fig. 11.19 Packets' statistics; conversation in Wireshark

Ethernet · 13	IPv4 · 281	IPv6 · 1	TCP · 1708	UDP · 841							
Address A	Address B	Packets	Bytes	Packets A→B	Bytes A→B	Packets B→A	Bytes B→A	Rel Start	Duration	Bits/s A→B	Bits/s B→A
00:0b:cd:c2:e4:91	01:00:5e:00:00:16	30	1860	30	1860	0	0	1.387136	727.0024	20	
00:0b:cd:c2:e4:91	ff:ff:ff:ff:ff:ff	6	749	6	749	0	0	6.547362	466.1334	12	
00:0b:cd:c2:e4:91	04:a1:51:ad:52:5a	24	2738	12	1754	12	984	6.547499	340.8939	41	
00:0b:cd:c2:e4:91	50:e5:49:e4:8b:d3	4	292	2	208	2	84	163.906844	308.7740	5	
01:00:5e:00:00:01	04:a1:51:ad:52:5a	29	1740	0	0	29	1740	21.044454	700.3667	0	
01:00:5e:00:00:16	04:a1:51:ad:52:5a	30	2100	0	0	30	2100	1.996130	723.6826	0	
01:00:5e:00:00:16	50:e5:49:e4:8b:d3	30	1620	0	0	30	1620	5.024018	725.6320	0	
04:a1:51:ad:52:5a	50:e5:49:e4:8b:d3	161,560	130 M	92,863	121 M	68,697	9315 k	0.000000	731.4960	1327 k	
04:a1:51:ad:52:5a	f0:76:1c:31:be:25	3	228	1	84	2	144	328.443071	0.0028	—	
04:a1:51:ad:52:5a	ff:ff:ff:ff:ff:ff	1	60	1	60	0	0	328.445624	0.0000	—	
04:a1:51:ad:52:5a	33:33:00:00:00:01	1	102	1	102	0	0	504.286065	0.0000	—	
50:e5:49:e4:8b:d3	f0:76:1c:31:be:25	8,941	8983 k	3,115	208 k	5,826	8774 k	18.317878	631.9327	2639	
50:e5:49:e4:8b:d3	ff:ff:ff:ff:ff:ff	3	469	3	469	0	0	163.906639	480.9379	7	

Fig. 11.20 Packets' statistics; conversation; name resolution in Wireshark

Ethernet · 18	IPv4 · 8	IPv6 · 1	TCP · 6	UDP · 7							
Address A	Address B	Packets	Bytes	Packets A→B	Bytes A→B	Packets B→A	Bytes B→A	Rel Start	Duration	Bits/s A→B	Bits/s B→A
00:17:33:61:00:00	e0:a1:d7:18:c2:73	321	171 k	161	149 k	160	22 k	1.794591	43.2607	27 k	
00:25:15:ae:ba:81	80:fb:06:f0:45:d7	1	60	0	0	1	60	5.216245	0.0000	—	
00:25:15:db:17:71	80:fb:06:f0:45:d7	1	60	0	0	1	60	20.218471	0.0000	—	
00:78:9e:5f:01:29	80:fb:06:f0:45:d7	1	60	0	0	1	60	35.228607	0.0000	—	
01:00:5e:7f:ff:fa	e0:a1:d7:18:c2:72	1	46	0	0	1	46	40.227545	0.0000	—	
24:95:04:05:b2:21	80:fb:06:f0:45:d7	1	60	0	0	1	60	0.219665	0.0000	—	
30:7e:cb:60:07:79	80:fb:06:f0:45:d7	1	60	0	0	1	60	25.214574	0.0000	—	
30:7e:cb:64:7f:e9	80:fb:06:f0:45:d7	1	60	0	0	1	60	20.217543	0.0000	—	
30:7e:cb:65:86:11	80:fb:06:f0:45:d7	1	60	0	0	1	60	35.226900	0.0000	—	
30:7e:cb:66:65:b1	80:fb:06:f0:45:d7	1	60	0	0	1	60	5.219666	0.0000	—	
30:7e:cb:79:8a:c9	80:fb:06:f0:45:d7	1	60	0	0	1	60	30.214086	0.0000	—	
30:7e:cb:81:98:61	80:fb:06:f0:45:d7	1	60	0	0	1	60	5.217729	0.0000	—	
30:7e:cb:8e:45:29	80:fb:06:f0:45:d7	1	60	0	0	1	60	0.222117	0.0000	—	
30:7e:cb:e2:29:09	80:fb:06:f0:45:d7	1	60	0	0	1	60	20.215783	0.0000	—	
64:7c:34:00:e0:9d	80:fb:06:f0:45:d7	1	60	0	0	1	60	0.221143	0.0000	—	
80:fb:06:f0:45:d7	e0:a1:d7:18:c2:72	10	1708	4	258	6	1450	0.000000	48.3301	42	
80:fb:06:f0:45:d7	e0:a1:d7:20:42:41	1	60	1	60	0	0	0.218216	0.0000	—	

Fig. 11.21 Packets' statistics; conversation in Wireshark for WiFi

The tool can extract certain forensic artifacts and extract them each to a separate folder in the output directory; Fig. 11.24.

The tool can also extract forensic artifacts from .PCAP network files. The link (http://www.behindthefirewalls.com/2014/01/extracting-files-from-network-traffic-PCAP.html) shows an example of how to use Foremost to extract forensic artifacts (e.g. executable and image files) from an input PCAP file.

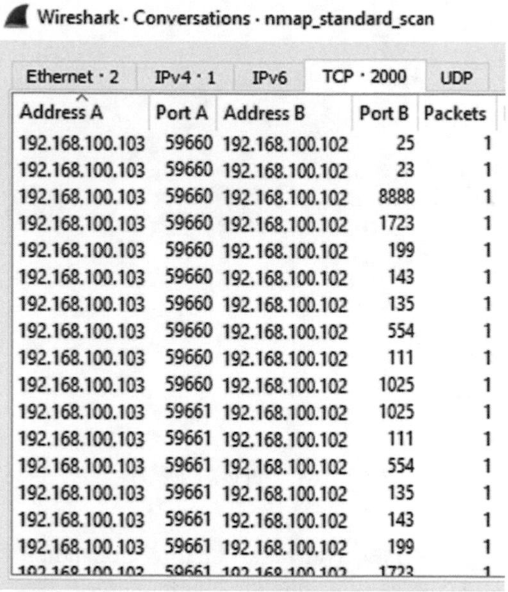

Fig. 11.22 Detection of
Port scanning using
Conversation tab.

Wireshark · Conversations · nmap_standard_scan

Ethernet · 2		IPv4 · 1		IPv6		TCP · 2000		UDP

Address A	Port A	Address B	Port B	Packets
192.168.100.103	59660	192.168.100.102	25	1
192.168.100.103	59660	192.168.100.102	23	1
192.168.100.103	59660	192.168.100.102	8888	1
192.168.100.103	59660	192.168.100.102	1723	1
192.168.100.103	59660	192.168.100.102	199	1
192.168.100.103	59660	192.168.100.102	143	1
192.168.100.103	59660	192.168.100.102	135	1
192.168.100.103	59660	192.168.100.102	554	1
192.168.100.103	59660	192.168.100.102	111	1
192.168.100.103	59660	192.168.100.102	1025	1
192.168.100.103	59661	192.168.100.102	1025	1
192.168.100.103	59661	192.168.100.102	111	1
192.168.100.103	59661	192.168.100.102	554	1
192.168.100.103	59661	192.168.100.102	135	1
192.168.100.103	59661	192.168.100.102	143	1
192.168.100.103	59661	192.168.100.102	199	1
192.168.100.102	59661	192.168.100.102	1723	1

Fig. 11.23 Foremost tool

Switches' Forensics

Switches connect machines at the local area network (LAN) level. They are considered as Layer 2 (L2) network devices. Switches uniquely distinguish packets and their originating or distance hosts through hosts MAC or physical addresses that operate in OSI layer 2. Traffic passing a switch can be of three different types based on intended receiver(s): Unicast which is intended to one or single receiver, multicast, intended for more than one receiver and finally broadcast intended to all switch ports or possible receivers.

They replace older (hubs) which were less "intelligent". Both use MAC or physical address to uniquely identify hosts in computer networks. Hubs broadcast traffic to the media of all local hosts. Switches minimize broadcast and pass traffic

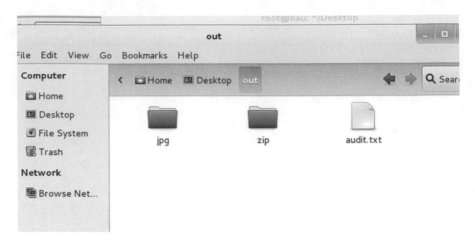

Fig. 11.24 Foremost tool output folder

to intended receiver rather than to all candidate receivers. You may see few hubs these days as switching are continuously taking their market shares. In some cases, some manufactures market a product as a hub while it works as a switch.

Many new, especially large switches can be managed and controlled through a web interfaces. Those switches are usually called: managed switches. Such switches are easier to interact with (e.g. for analysis and monitoring) in comparison with classical switches that can be only accessed, for control and management, by only special cables and using their own console or terminal. Additionally, there are other types of switches such as "Layer 3" switches which can have routing in addition to switching functionality.

Typically, the first layer of connections in any network or enterprise includes switches (i.e. edge switches). Then based on the size of the network, the second layer can be also switches (i.e. core switches) until the last layer that connects with remote hosts or the Internet which will then include routers rather than switches.

Switches include "switching or CAM tables" where each MAC address will have port/interface name/number. This means that a traffic destined to this MAC or physical address will be sent through related port or interface. Those roles in the switching tables are manually written by network administrators. In most recent types of switches: Software Defined Networking or programmable switches, a proactive mode can be defined in switches. In this mode, the switch will dynamically add those flow-roles dynamically based on incoming and outgoing traffic.

Attackers may use different techniques to monitor switch traffics. For example, they may flood the switch with too many traffic. Their goal is to saturate CAM tables with many records. Once switch CAM table is saturated and can't receive any more records, it will reach a state "Fail open" in which the switch will broadcast all incoming traffic.

ARP spoofing (or cache poisoning) can be also used by attackers to gain access to switch CAM tables. Address Resolution Protocol (ARP) is used by switches to

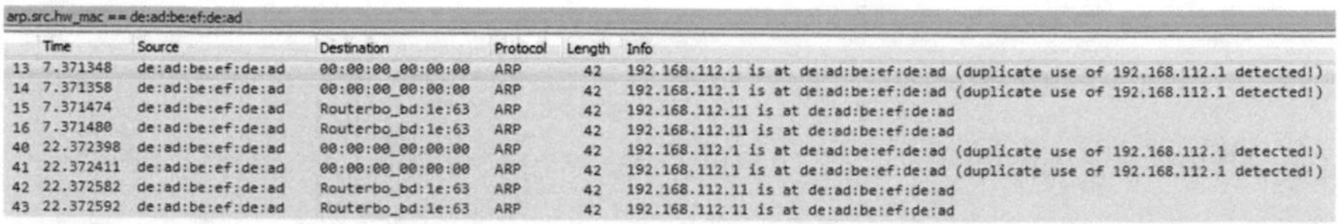

```
arp.src.hw_mac == de:ad:be:ef:de:ad
```

Time	Source	Destination	Protocol	Length	Info
13 7.371348	de:ad:be:ef:de:ad	00:00:00_00:00:00	ARP	42	192.168.112.1 is at de:ad:be:ef:de:ad (duplicate use of 192.168.112.1 detected!)
14 7.371358	de:ad:be:ef:de:ad	00:00:00_00:00:00	ARP	42	192.168.112.1 is at de:ad:be:ef:de:ad (duplicate use of 192.168.112.1 detected!)
15 7.371474	de:ad:be:ef:de:ad	Routerbo_bd:1e:63	ARP	42	192.168.112.11 is at de:ad:be:ef:de:ad
16 7.371480	de:ad:be:ef:de:ad	Routerbo_bd:1e:63	ARP	42	192.168.112.11 is at de:ad:be:ef:de:ad
40 22.372398	de:ad:be:ef:de:ad	00:00:00_00:00:00	ARP	42	192.168.112.1 is at de:ad:be:ef:de:ad (duplicate use of 192.168.112.1 detected!)
41 22.372411	de:ad:be:ef:de:ad	00:00:00_00:00:00	ARP	42	192.168.112.1 is at de:ad:be:ef:de:ad (duplicate use of 192.168.112.1 detected!)
42 22.372582	de:ad:be:ef:de:ad	Routerbo_bd:1e:63	ARP	42	192.168.112.11 is at de:ad:be:ef:de:ad
43 22.372592	de:ad:be:ef:de:ad	Routerbo_bd:1e:63	ARP	42	192.168.112.11 is at de:ad:be:ef:de:ad

Fig. 11.25 An example of ARP spoofing

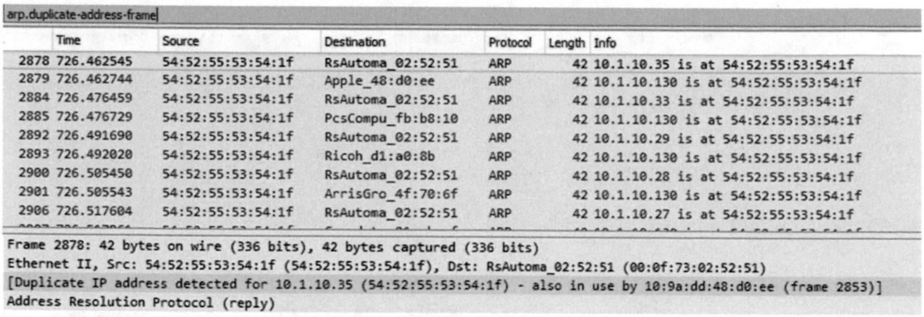

```
arp.duplicate-address-frame
```

Time	Source	Destination	Protocol	Length	Info
2878 726.462545	54:52:55:53:54:1f	RsAutoma_02:52:51	ARP	42	10.1.10.35 is at 54:52:55:53:54:1f
2879 726.462744	54:52:55:53:54:1f	Apple_48:d0:ee	ARP	42	10.1.10.130 is at 54:52:55:53:54:1f
2884 726.476459	54:52:55:53:54:1f	RsAutoma_02:52:51	ARP	42	10.1.10.33 is at 54:52:55:53:54:1f
2885 726.476729	54:52:55:53:54:1f	PcsCompu_fb:b8:10	ARP	42	10.1.10.130 is at 54:52:55:53:54:1f
2892 726.491690	54:52:55:53:54:1f	RsAutoma_02:52:51	ARP	42	10.1.10.29 is at 54:52:55:53:54:1f
2893 726.492020	54:52:55:53:54:1f	Ricoh_d1:a0:8b	ARP	42	10.1.10.130 is at 54:52:55:53:54:1f
2900 726.505450	54:52:55:53:54:1f	RsAutoma_02:52:51	ARP	42	10.1.10.28 is at 54:52:55:53:54:1f
2901 726.505543	54:52:55:53:54:1f	ArrisGro_4f:70:6f	ARP	42	10.1.10.130 is at 54:52:55:53:54:1f
2906 726.517604	54:52:55:53:54:1f	RsAutoma_02:52:51	ARP	42	10.1.10.27 is at 54:52:55:53:54:1f

```
Frame 2878: 42 bytes on wire (336 bits), 42 bytes captured (336 bits)
Ethernet II, Src: 54:52:55:53:54:1f (54:52:55:53:54:1f), Dst: RsAutoma_02:52:51 (00:0f:73:02:52:51)
[Duplicate IP address detected for 10.1.10.35 (54:52:55:53:54:1f) - also in use by 10:9a:dd:48:d0:ee (frame 2853)]
Address Resolution Protocol (reply)
```

Fig. 11.26 Using a filter to detect ARP spoofing

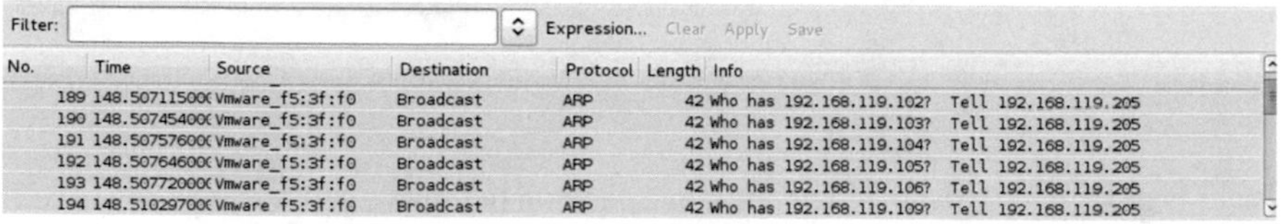

| Filter: | | ⬍ | Expression... | Clear | Apply | Save |

No.	Time	Source	Destination	Protocol	Length	Info
189 148.50711500	Vmware_f5:3f:f0	Broadcast	ARP	42	Who has 192.168.119.102? Tell 192.168.119.205
190 148.50745400	Vmware_f5:3f:f0	Broadcast	ARP	42	Who has 192.168.119.103? Tell 192.168.119.205
191 148.50757600	Vmware_f5:3f:f0	Broadcast	ARP	42	Who has 192.168.119.104? Tell 192.168.119.205
192 148.50764600	Vmware_f5:3f:f0	Broadcast	ARP	42	Who has 192.168.119.105? Tell 192.168.119.205
193 148.50772000	Vmware_f5:3f:f0	Broadcast	ARP	42	Who has 192.168.119.106? Tell 192.168.119.205
194 148.51029700	Vmware_f5:3f:f0	Broadcast	ARP	42	Who has 192.168.119.109? Tell 192.168.119.205

Fig. 11.27 Port scanning activities

map MAC addresses to IP addresses. Each record in the ARP table includes the logical IP addresses associated with the physical MAC address. An attacker will try to add or edit a record in this table to legitimize their host and make it look like a local host. We can use Wireshark to detect a possible ARP spoofing (Fig. 11.25). The Figure shows that two different IP addresses have the same MAC or physical address. Wireshark warned against that; by showing a duplicate use message. You can use a filter such as: (arp.duplicate-address-frame) to detect ARP spoofing in a packet stream; Fig. 11.26.

We may need to check if there was "port scanning activities", using for example Nmap. Port scanners use ARP (for local networks) and PING (for external networks) protocols to scan network or system ports. Figure 11.27 shows an indication of port scanning in a local network. Port scanning is used usually in first stages of network attacks.

In Windows environments, Nmap typically uses TCP port 135 (Fig. 11.28).

If we want to detect a DHCP server, we can use a filter such as: (bootp.option. type=53). Based on following packets in the Fig. 12.18, we can see the DHCP

| Filter: | tcp.port==135 | | | ⌄ | Expression... | Clear | Apply | Save |

No.	Time	Source	Destination	Protocol	Length	Info
4023	601.0664	192.168.119.205	192.168.119.219	TCP	74	43703 > 135 [SYN] Seq=0 W
4026	601.0668	192.168.119.219	192.168.119.205	TCP	70	135 > 43703 [SYN, ACK] Se
4027	601.0668	192.168.119.205	192.168.119.219	TCP	66	43703 > 135 [ACK] Seq=1 A
4028	601.0669	192.168.119.205	192.168.119.219	TCP	66	43703 > 135 [RST, ACK] Se

Fig. 11.28 Filter using TCP port 135 for Port scanning

| bootp.option.type == 53 | | | | | | | ☒ ☐ ▾ |

No.	Time	Source	Destination	Protocol	Length	Info
1	0.000000	0.0.0.0	255.255.255...	DHCP		342 DHCP Discover - Transaction ID 0x61d6b369
2	0.000416	192.168.40.1	192.168.40.10	DHCP		342 DHCP Offer - Transaction ID 0x61d6b369
3	0.000812	0.0.0.0	255.255.255...	DHCP		368 DHCP Request - Transaction ID 0x61d6b369
4	0.001211	192.168.40.1	192.168.40.10	DHCP		342 DHCP ACK - Transaction ID 0x61d6b369
5	2.563151	192.168.40.10	192.168.40.1	DHCP		356 DHCP Request - Transaction ID 0x1f696075
6	2.563579	192.168.40.1	192.168.40.10	DHCP		342 DHCP ACK - Transaction ID 0x1f696075

Fig. 11.29 A query for DHCP

| tcp.flags.syn==1 && tcp.flags.ack==0 | | | | | | ☒ ☐ ▾ Expression... |

	Time	Source	Destination	Protocol	Length	Info
11	15.562275	192.168.40.10	php.net	TCP		62 nsstp(1036) → http(80) [SYN] Seq=0 Win=65535 Len=0 _
28	15.616110	192.168.40.10	y3b.php.net	TCP		62 aws(1037) → http(80) [SYN] Seq=0 Win=65535 Len=0 MS_
29	15.616686	192.168.40.10	y3b.php.net	TCP		62 xtqp(1038) → http(80) [SYN] Seq=0 Win=65535 Len=0 M_
34	15.619447	192.168.40.10	y3b.php.net	TCP		62 sbl(1039) → http(80) [SYN] Seq=0 Win=65535 Len=0 MS_
88	16.042142	192.168.40.10	url.whichusb_	TCP		62 netarx(1040) → http(80) [SYN] Seq=0 Win=65535 Len=0_
152	17.669844	192.168.40.10	aes.whichdig_	TCP		62 danf-ak2(1041) → http(80) [SYN] Seq=0 Win=65535 Len_
160	18.605511	192.168.40.10	zivvgwyrwy.3_	TCP		62 afrog(1042) → http(80) [SYN] Seq=0 Win=65535 Len=0 _
173	19.945752	192.168.40.10	zivvgwyrwy.3_	TCP		62 boinc-client(1043) → http(80) [SYN] Seq=0 Win=65535_
174	19.953791	192.168.40.10	zivvgwyrwy.3_	TCP		62 dcutility(1044) → http(80) [SYN] Seq=0 Win=65535 Le_

Fig. 11.30 A filter to detect TCP requests without acknowledgements

server is at IP: 192.168.40.1. The first four packets belong to one transaction (based on the transaction ID) and the last two belong to another one (Fig. 11.29).

Figure 11.30 shows a possible filter to be used to detect SYN flooding attacks. When a system is flooded with too many connection requests, it will not be able to provide timely acknowledgment ACK messages (tcp.flags.syn == 1 && tcp.flags.ack == 0).

Different forensic cases may trigger different triggers. In Fig. 11.31, a filter is used to follow-up on an attack which used FTP protocols. The Figure shows a filter to detect user name or password that was used in an attack. Packet capture file used in the analysis can be downloaded from: https://github.com/markofu/pcaps.

Port Mirroring

In port mirroring, packets, entering a port are copied where this copy is sent for a special interface for monitoring (local or remote). Many traffic analysis activities use this copied traffic for traffic monitoring and security issues. This task of copying traffic (i.e. the amount and levels of traffic to copy) is usually a result of a balance

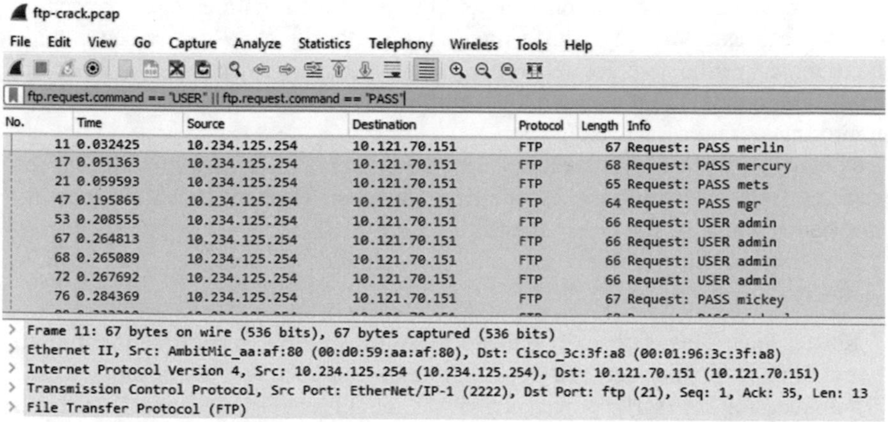

Fig. 11.31 A filter to detect user name or password used in an attack

between security issues (which requires more quality and quantity of traffic copying) and performance, which requires less. Port mirroring can be conFigured to mirror only traffic that meets certain criteria (i.e. coming from or going to a certain interface, VLAN, port, etc.). Roles can be sent from a firewall (e.g. action: port-mirror) that commands some traffic to be mirrored. Port mirroring in Cisco systems is usually referred to as SPAN (Switched Port Analyzer).

Port mirroring process faces many practical challenges. A switch may typically have 1, 2, or 3 mirror ports. It will be impractical, from a bandwidth perspective, for those few ports to receive traffic from all other ports in the switch (which may have 24, 48 ports, etc.).

Routers' Forensics

Routers, in comparison with switches, are typically Wide Area Networks (WAN) rather than LAN components. They are smarter and more complex than switches. They can also be used decompose a large networks into sub-networks, with different addressing schemes, and connect different subnets of the network together. Similar to switching tables, routers have routing tables with IP addresses and destinations. However, routing decisions are more complex than switching decisions as the candidate destinations are much larger. Each router uses one or more routing algorithm to make decisions on routes for the different packets. Typically, distance, number of hubs and quality/reliability of those hubs in the path from source to destination are all factors, and many others, that are considered when making routing decisions.

A forensic investigator may try to reverse engineer the path that some traffic traversed in the past based on routing algorithms.

Routers can also include many other network functionalities such as: access controls (e.g. as a firewall), packet filtering, monitoring, etc. Routers can include one or more specific ports for network traffic monitoring (i.e. span ports). They can also be programmed to route such logs or information to certain destinations (e.g. a monitoring system, server, etc.).

Routers have their own memory, like computers. Many information can be extracted from evaluating their memories. However, we may lose such memory information when the router is rebooted.

Routers can have several artifacts that should be searched for possible evidences:

- **Router logs:** This is typically the largest source of forensic information in routers. Similar to other network devices, routers can keep records of inbound and outbound traffic. Traffic or packet analysis includes the same network attributes that can be seen in other network devices. This means that the same techniques we explained it in previous sections can be used to find forensic artifacts for router logs. The level of details that can be found in router logs can be different based on three factors:

    - **Router manufacture:** Different manufactures may have some unique log contents that can be found in router logs and can be used to uniquely identify the router and its manufacture.
    - **Router function:** Routers can be used for many different functions. They can be very small, serving a home user; wireless access points or routers. They can be also in companies with small, medium to large size enterprises. Routers can be also used as WAN equipments to connect different enterprises, companies, or internet sections with each other. The level of details and the nature of logged information can be different based on those different functions.
    - **Logging levels:** In many cases, network administrators can make decisions on what or what not to log from network activities. This can be typically a compromise between security and performance.

- **Router memory:** Similar to computing machine or operating systems, routers have a volatile memory. This memory contains router configurations, any additions or changes to those configurations. In many forensic cases that involve malwares or hacking, it is possible that intruders changed some of the router configurations to help them in their attack. Traces of those edits or changes can be searched for in router memory. Memory forensic tools that were described in other chapters can be used to analyze routers' memories.

- **Routing protocols and routing tables:** Routing protocols can be extracted from routers configuration or memory. As we mentioned earlier in some forensic cases related to the analysis of malwares or hacking attacks, it is important to check routing protocols and routing tables to make sure that those were not tampered or changed as a result of the malwares or the attacks.

Firewalls' Forensics

Classically, firewalls are not meant to be complicated. Performance and robustness qualities are very important in firewalls. Hence to ensure those quality attributes, logging is not meant to be part of firewall tasks. Firewalls are action points to decide what to do with traffic or access requests (simply permit or deny). Nonetheless, firewalls can be conFigured to log and audit traffic and access request activities. Such logging or auditing is not recommended all the time as it will significantly impact performance, especially for network-based or system-wide firewalls. Firewalls can exist on each computer to monitor/control traffic on that specific computer or host; host-based firewalls. Firewalls can also exist in a network to monitor/control traffic for the whole network; network-based firewalls.

The most two important artifacts that a network forensic is interested to look in a firewall are the firewall roles and logs:

- **Firewall roles:** Firewalls roles can be analyzed as part of understanding network traffic and activities. Investigators should quickly scan those roles for any existence of odd or abnormal roles. Then they should look for certain roles based on the forensic case context. For example, if this is a case related to a malware that used a form of Denial of Service (DoS) attack, we may look for roles that could have possibly blocked it (e.g. based on payload, packet size, application port number, etc.). Those can be also used for post-mortem activities to add/ modify some roles to block such malwares in future. While such task is not part of the forensic investigation, however, many forensic investigations are not only conducted for legal issues but also to detect and fix security problems.
- **Firewall logs:** As we mentioned earlier, classically, firewalls are not meant to log and audit network activities. They are meant to be action points and take actions on inbound and outboard traffic. This was extremely important to keep firewall performance and robustness high. However, more recently firewall goals are expanded and somewhat integrated with any malware systems as well as IDS/ IPS. Those firewalls include logs of the different activities or actions that the firewall took. Such logs however are not comprehensive and may not represent the full picture or view of network traffic. It also depends on the nature and quality of firewalls roles in the firewall. Those roles are created and maintained manually by system or network administrators.

IDS/IPS Forensics

Intrusion Detection/Protection Systems (IDS/IPS) can be considered as the most complex and intelligent network devices. In comparison with firewalls IDS/IPS network components take more thorough decisions. Firewalls are meant to be primitive and take decisions based on primary traffic attributes (i.e. source and destination:

IP addresses, MAC addresses, port numbers in addition to protocols). In addition to this, IDS/IPS can take actions (i.e. permit, deny, alert, etc.) based on deep packets inspections or based on some behaviors that require investigation more than a single packet.

If compared with firewalls, IDS/IPS design and implementation can widely vary. Firewalls have standard procedures and actions regardless of the company originating or making that firewall. On the other hand, IDS/IPS design and the nature and complexity of the functions they can do can widely vary from one system to another. IDS/IPS roles can be modified/ changed to fit a particular ongoing investigation. For example, forensic investigators may request IDS/IPS to add a temporary role to log packets that match a certain patter (e.g. a pattern that is crafted to match a suspect profile).

SNORT (https://www.snort.org) is a popular example of an IDS/IPS open source system. It evolves through its history with more complexity and features. It has been also recently acquired as a commercial product, although users can still download an open source or free version. A large inventory of rules (https://www.snort.org/rules_explanation) for different categories of functionalities or attacks can be also downloaded. Users can choose to enable certain rules. They can also modify such rules or create their own rules. See skills/experience section on exercises related to SNORT.

Many current system defense applications combine roles of: Anti-viruses, firewalls and IDS/IPS. They can alert, permit or deny access requests from applications (i.e. layer7). They can also do that as firewalls (i.e. layers 2 and 3). They can also control traffic based on complex roles or certain behaviors to act as IDS/IPS. IDS/IPSs can be defined to control packets based on:

- Signatures: There are many malwares who are known. They have been analyzed to detect how to, uniquely distinguish such malwares. This is the main and general detection method in IDS/IPS. They can be considered as an extension to firewalls as firewalls can detect malwares with signatures related to basic network information (i.e. source or destination IP, MAC addresses or port numbers, protocols, etc.). IDS/IPS can extend this signature-based detection to information that can be extracted from packet headers, size of packets, payloads, etc. The process is simple as it usually employs *simple* value to value comparison between values in the subject or tested packet to the values recorded for known malwares.
- Behavioral: This is much more complex that signature-based detection or analysis. The approach should first record what can be considered as "normal traffic behavior" and then alert against any deviations from such normal behavior. Clearly, there are many challenges in this approach. First, the definition of what is normal or abnormal can be very complex to accommodate. This means that we may have many cases of false positives or false negatives. Those are incorrect detections by IDS/IPS detection malicious traffic as normal traffic; false negative, or detecting normal traffic as malicious; false positive.
- Protocol conformance: Many malwares try to change some of the specifications of applications or protocols against their standards (e.g. RFCs). An IDS/IPS that

can track all those standards can check if the current packet includes deviations from those standards. One challenge in this aspect is that many manufactures try to have their own implementations of some applications or protocols that may deviate slightly from defined standards.

In IDS/IPS there are typically two main approaches for forensic analysts. In malware analysis in particular, forensic analysts can instruct IDS/IPS to monitor or alert for traffic that forensic investigators can define (i.e. roles, signature, behavior or conformance based). This is usually a proactive approach with the goal of trying to detect or analyze malwares.

The second approach is through using IDS/IPS logs for evidence acquisition. Similar to many other network devices or controls, IDS/IPS can create logs of activities, if they are instructed to do so. Those logs can be analyzed searching for possible relevant evidences to the case in-hand. The evidence can be a payload (e.g. a file, a trace of action, corrupting of data, etc.). IDS/IPS logs may contain the evidence or may lead to it.

Wireless Forensics

Wireless communication is continuously growing as it is more convenient in comparison with wired communication. It is growing in both dimensions, in width (i.e. different types of wireless communication media) and depth or volume with more and more number of users. In this scope, wireless networks can include:

- Mobile or smart phones. As this is growing to be very important and large, we dedicated a chapter for mobile forensics in this book and hence this will not be covered in this chapter.
- WiFi networks. The current percentage of computers, laptops, tablets and smart phones that are using the Internet using WiFi in comparison with wired (i.e. fiber optic, twisted pairs, cable modems and twisted pairs) is large and continuously growing. Eventually, wired connections will be strictly to servers, and some business machines that are largely stable in their location. This is synchronized with the growing number of laptops, tablets, and smart phones in comparison with desktops and servers.
- Bluetooth connections: Used in short distances for different types of applications.
- Infrared: Infrared wireless is used in TV, and remote controls from many devices.
- Cordless phones

From practical experience, most focus will be in this section on WiFi networks. Other wireless media have less applicability of forensic cases, and less availability of forensic tools.

For many network analysis tools, wireless is just another interface. When we described Wireshark earlier, we showed how Wireshark can be constructed to listen

No.	Time	Source	Destination	Protocol	Length	Info
1	0.000000	192.168.1.132	192.168.1.1	DNS	181	Standard query 0x96c1 A www.polito.it
2	0.000020		GemtekTe_cb:6e:1a (…	802.11	46	Acknowledgement, Flags=........C
3	0.000036	192.168.1.1	192.168.1.132	DNS	174	Standard query response 0x96c1 A www.|
4	0.000052		GemtekTe_cd:74:7b (…	802.11	46	Acknowledgement, Flags=........C
5	0.001716	192.168.1.132	130.192.73.1	TCP	186	3827 → 80 [SYN] Seq=0 Win=65535 Len=0
6	0.001733		GemtekTe_cb:6e:1a (…	802.11	46	Acknowledgement, Flags=........C
7	0.203162	130.192.73.1	192.168.1.132	TCP	122	80 → 3827 [SYN, ACK] Seq=0 Ack=1 Win=
8	0.203180		GemtekTe_cd:74:7b (…	802.11	46	Acknowledgement, Flags=........C
9	0.203195	192.168.1.132	130.192.73.1	TCP	162	3827 → 80 [ACK] Seq=1 Ack=1 Win=26280

> Frame 1: 181 bytes on wire (1448 bits), 181 bytes captured (1448 bits)
> PPI version 0, 84 bytes
> 802.11 radio information
> IEEE 802.11 QoS Data, Flags: ....TC
> Logical-Link Control
> Internet Protocol Version 4, Src: 192.168.1.132, Dst: 192.168.1.1
> User Datagram Protocol, Src Port: 1031, Dst Port: 53

Fig. 11.32 An example of WiFi packets

to different types of interfaces including WiFi or Bluetooth interfaces. A wireless access point, or hotspot, can act as a hub, switch or router. This depends on the detail functionalities of the access point and whether it is used for LAN, WAN or Internet services. While computers connected to an access point will not be using a specific port to connect to the access point, yet they will have unique IP and MAC addresses upon which they can be identified. The physical layer of a wireless spectrum has 11 different channels. The investigated traffic should be seen in one of those 11 channels. Packet loss data is possible based on the nature of surrounding obstacles. It is also possible to sniff packets from different WiFi access points, if the distance between those different access points is short.

WiFi can be connected in an infrastructure mode through an access point. They can be also connected to each other; peer to peer in an ad-hoc network.

Different tools can be used to capture traffic from WiFi networks. Airpcap, which is integrated with Wireshark can be used to do that. In analyzing WiFi packets, we can new sections related to (802.11); Fig. 12.21. Samples of WiFi packets can be found in Wireshark website (https://wiki.wireshark.org/SampleCaptures). In order to correlate IP/MAC addresses with hosts, simple commands such as: ipconfig, ifconfig or iwconfig can be used. Those will help identify the basic network information and interfaces for each host or computer. However, in a DHPC environment, current IP addresses can be different from those allocated at the time of the incident or when the PCAP file was extracted (Fig. 11.32).

In this sample of WiFi packets, if we want to check for unencrypted data, we can use the filter (wlan.fc.protected == 0 && wlan.fc.type eq 2); Fig. 11.33.

Table 11.1 shows the different filters that can be used and the frames that they will retrieve.

If we know our target WiFi network (i.e. SSID), we can filter traffic to only see packets from this target network, Fig. 11.34.

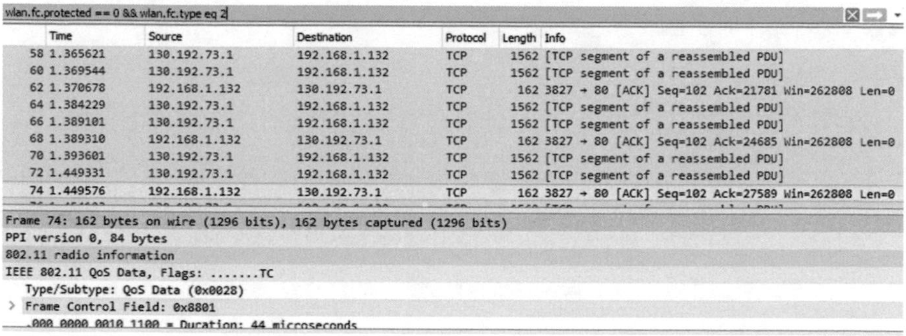

```
wlan.fc.protected == 0 && wlan.fc.type eq 2                                                    ☒ ⊑ ▾
      Time         Source              Destination         Protocol  Length  Info
  58 1.365621      130.192.73.1        192.168.1.132       TCP       1562 [TCP segment of a reassembled PDU]
  60 1.369544      130.192.73.1        192.168.1.132       TCP       1562 [TCP segment of a reassembled PDU]
  62 1.370678      192.168.1.132       130.192.73.1        TCP        162 3827 → 80 [ACK] Seq=102 Ack=21781 Win=262808 Len=0
  64 1.384229      130.192.73.1        192.168.1.132       TCP       1562 [TCP segment of a reassembled PDU]
  66 1.389101      130.192.73.1        192.168.1.132       TCP       1562 [TCP segment of a reassembled PDU]
  68 1.389310      192.168.1.132       130.192.73.1        TCP        162 3827 → 80 [ACK] Seq=102 Ack=24685 Win=262808 Len=0
  70 1.393601      130.192.73.1        192.168.1.132       TCP       1562 [TCP segment of a reassembled PDU]
  72 1.449331      130.192.73.1        192.168.1.132       TCP       1562 [TCP segment of a reassembled PDU]
  74 1.449576      192.168.1.132       130.192.73.1        TCP        162 3827 → 80 [ACK] Seq=102 Ack=27589 Win=262808 Len=0
Frame 74: 162 bytes on wire (1296 bits), 162 bytes captured (1296 bits)
PPI version 0, 84 bytes
802.11 radio information
IEEE 802.11 QoS Data, Flags: .......TC
   Type/Subtype: QoS Data (0x0028)
 > Frame Control Field: 0x8801
   .000 0000 0010 1100 = Duration: 44 microseconds
```

Fig. 11.33 A filter to check for unencrypted data

Table 11.1 Examples of WiFi Wireshark filters

Frame Type/Subtype	Filter Syntax
Management frames	wlan.fc.type eq 0
Control frames	wlan.fc.type eq 1
Data frames	wlan.fc.type eq 2
Association request	wlan.fc.type_subtype eq 0
Association response	wlan.fc.type_subtype eq 1
Reassociation request	wlan.fc.type_subtype eq 2
Reassociation response	wlan.fc.type_subtype eq 3
Probe request	wlan.fc.type_subtype eq 4
Probe response	wlan.fc.type_subtype eq 5
Beacon	wlan.fc.type_subtype eq 8
Disassociate	wlan.fc.type_subtype eq 10
Authentication	wlan.fc.type_subtype eq 11
Deauthentication	wlan.fc.type_subtype eq 12
Action frames	wlan.fc.type_subtype eq 13
Block ACK requests	wlan.fc.type_subtype eq 24
Block ACK	wlan.fc.type_subtype eq 25
Power save poll	wlan.fc.type_subtype eq 26
Request to send	wlan.fc.type_subtype eq 27
Clear to send	wlan.fc.type_subtype eq 28
ACK	wlan.fc.type_subtype eq 29
Contention free period end	wlan.fc.type_subtype eq 30
NULL data	wlan.fc.type_subtype eq 36
QoS data	wlan.fc.type_subtype eq 40
Null QoS data	wlan.fc.type_subtype eq 44

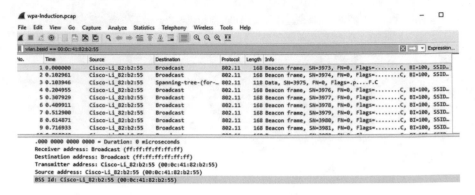

Fig. 11.34 Filter traffic based on WiFi SSID

Skills' Sections

Traffic Analysis

- Using the link (https://uscc.cyberquests.org/), pick 1 year competition and download traffic file. Then answer all questions in the selected competition based on analyzing included traffic files.
- Using the link (http://forensicscontest.com/), pick 1 year competition and download traffic file. Then answer all questions in the selected competition based on analyzing included traffic files.
- In the link (http://honeynet.org/node/122), a detail description and solution for a network forensic challenge (The Honeynet Project). Download the file (conference. PCAPng.gz) from the link above. Then use Wireshark (or any proper tool) to open and analyze the file to show the case that was described in the link. In each case describe the packet(s) with their numbers that are related to the different evidences.
- Download again the dataset from: http://www.cs.technion.ac.il/~gnakibly/ TCPInjections/samples.zip. Then for each row in this table (which includes an injection event), create a new row and take only the first column from their table as the connection to your table, then add the following new columns/ fields (file(s): [1] which files in the dataset describe the event, [2] event type: rouge advertisement (8 instances), malicious intent (4 instances), block content (2 instances), [3] which packets show the event and [4] your own description of the event.

Group name	Files(s)	Event type	Packets to show event	Description/ screenshots, etc

Foremost

Using details about how to use Foremost in this chapter, show how Foremost tool can be used to analyze a network PCAP file. Select any PCAP file from open source references listed in this chapter and elsewhere to demo this and what forensic artifacts can be extracted from the input PCAP file.

ARP Spoofing

At least one of the PCAP files in (http://uscc.cyberquests.org/assets/cyberquest_ spring2017.zip) has an instance of ARP spoofing. Using Wireshark, find this ARP spoofing case and show how you were able to detect or discover it.

Applications' Section

NIC Promisc Mode

Make your own research to show: How to find out that an NIC is in promiscuous mode on a LAN? Demo the tool or process and show how you can enable\disable an NIC to promisc mode then detect that it is running in a promisc mode.
   Hint: Some possible approaches:

- Based on making DNS queries or name lookups
- Based on testing with ARP and broadcasting message.
- Looking at the differences in PING response time.
- There are some tools/scripts that can be used (e.g. SniffDet (http://sniffdet. sourceforge.net/), NMAP (http://nmap.org/nsedoc/scripts/sniffer-detect.html), NAST (http://nast.berlios.de/), PTool (https://code.google.com/p/ptool/), Cain and Abel (http://www.oxid.it/ca_um/topics/promiscuous-mode_scanner.htm, sniffing tab), Promqry (http://support.microsoft.com/kb/892853),

SDN Switches

Download a version of SDN/OpenFlow or programmable switches. Show how such switch can be used in proactive or dynamic mode where roles are added dynamically based on incoming/outgoing traffic. You can try examples of some of the open source SDN controllers/switches: OpenDayLight, https://www.opendaylight.org, Floodlight: www.projectfloodlight.org/floodlight/, Ryu: https://osrg.github.io/ryu, POX: https://github.com/noxrepo/pox, etc.

Switch Forensics

Refer to a tool that we developed to analyze switch products: https://github.com/
alsmadi/SDN-Competition. You are expected to correlate information from switch
memory dump and packet files. Download the 2 files from: https://www.cmand.org/
sdn/sdnf.html. Then try to answer questions posted in the same page based on information from the two files.

IDS/IPS Forensics

We will be using SNORT NIDS in this assignment. It's easy to use it or install it
within Linux based systems. If you have one, you dont need to use Deterlab for this
assignment. If you dont have one, reserve one machine in Deterlab and use it for this
assignment (Reserve one or two machines)

- You can use the NS file below (to reserve 2 nodes)

  -------------------------------- SNORT.ns------------------------------------

```
set ns [new Simulator]
sourcetb_compat.tcl
set node1 [$ns node]
set node2 [$ns node]
tb-set-node-os $node1 Ubuntu1204-64-STD
tb-set-node-os $node2 Ubuntu1204-64-STD
#tb-set-node-memory-size $node1 512
#tb-set-node-memory-size $node2 512
#tb-set-node-startcmd node0 startupcmd
set link0 [$ns duplex-link $node1 $node2 100000.0kb 0.0ms DropTail]
$ns rtproto Static
$ns run
```

- Step 1: Install and conFigure snort
- **$sudo apt-get install snort**
- Read the comments in **/etc/snort/snort.conf** carefully and pay attention to the
  definition of variable HOME_NET and EXTERNAL_NET.

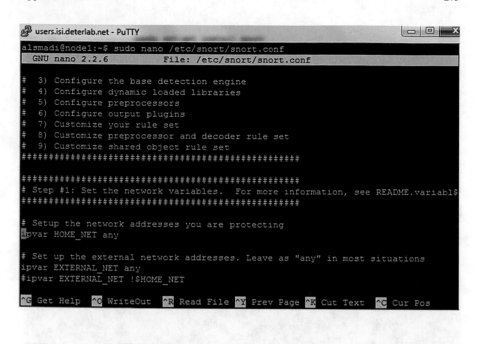

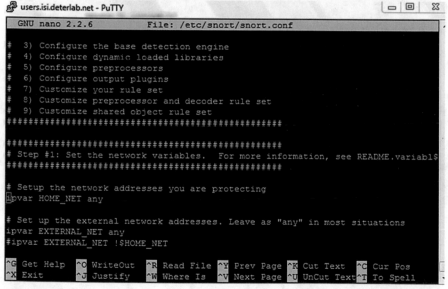

- You can test configuration by:

- `sudosnort -T -c /etc/snort/snort.conf`

```
alsmadi@node1:~$ sudo nano /etc/snort/rules/local.rules
alsmadi@node1:~$ sudo /etc/init.d/snort restart[]
```

```
users.isi.deterlab.net - PuTTY                                    [_][□][X]

       Copyright (C) 1998-2011 Sourcefire, Inc., et al.
       Using libpcap version 1.1.1
       Using PCRE version: 8.12 2011-01-15
       Using ZLIB version: 1.2.3.4

       Rules Engine: SF_SNORT_DETECTION_ENGINE  Version 1.15  <Build 18>
       Preprocessor Object: SF_FTPTELNET (IPV6)  Version 1.2  <Build 13>
       Preprocessor Object: SF_SIP (IPV6)  Version 1.1  <Build 1>
       Preprocessor Object: SF_SSLPP (IPV6)  Version 1.1  <Build 4>
       Preprocessor Object: SF_SMTP (IPV6)  Version 1.1  <Build 9>
       Preprocessor Object: SF_MODBUS (IPV6)  Version 1.1  <Build 1>
       Preprocessor Object: SF_REPUTATION (IPV6)  Version 1.1  <Build 1>
       Preprocessor Object: SF_DNS (IPV6)  Version 1.1  <Build 4>
       Preprocessor Object: SF_DCERPC2 (IPV6)  Version 1.0  <Build 3>
       Preprocessor Object: SF_DNP3 (IPV6)  Version 1.1  <Build 1>
       Preprocessor Object: SF_POP (IPV6)  Version 1.0  <Build 1>
       Preprocessor Object: SF_SSH (IPV6)  Version 1.1  <Build 3>
       Preprocessor Object: SF_GTP (IPV6)  Version 1.1  <Build 1>
       Preprocessor Object: SF_SDF (IPV6)  Version 1.1  <Build 1>
       Preprocessor Object: SF_IMAP (IPV6)  Version 1.0  <Build 1>

Snort successfully validated the configuration!
Snort exiting
alsmadi@node1:~$ []
```

- Then try running Snort as root: (basic command with no options sudo snort, monitoring mode). Image below after trying ping to the node.

```
users.isi.deterlab.net - PuTTY                                    [_][□][X]

Type:0  Code:0  ID:19257  Seq:170  ECHO REPLY
=+=+=+=+=+=+=+=+=+=+=+=+=+=+=+=+=+=+=+=+=+=+=+=+=+=+=+=+=+=+=+=+

10/10-18:51:28.335221 10.1.1.3 -> 10.1.1.2
ICMP TTL:64 TOS:0x0 ID:60290 IpLen:20 DgmLen:84 DF
Type:8  Code:0  ID:19257  Seq:171  ECHO
=+=+=+=+=+=+=+=+=+=+=+=+=+=+=+=+=+=+=+=+=+=+=+=+=+=+=+=+=+=+=+=+

10/10-18:51:28.335249 10.1.1.2 -> 10.1.1.3
ICMP TTL:64 TOS:0x0 ID:62648 IpLen:20 DgmLen:84
Type:0  Code:0  ID:19257  Seq:171  ECHO REPLY
=+=+=+=+=+=+=+=+=+=+=+=+=+=+=+=+=+=+=+=+=+=+=+=+=+=+=+=+=+=+=+=+

10/10-18:51:29.335128 10.1.1.3 -> 10.1.1.2
ICMP TTL:64 TOS:0x0 ID:60483 IpLen:20 DgmLen:84 DF
Type:8  Code:0  ID:19257  Seq:172  ECHO
=+=+=+=+=+=+=+=+=+=+=+=+=+=+=+=+=+=+=+=+=+=+=+=+=+=+=+=+=+=+=+=+

10/10-18:51:29.335158 10.1.1.2 -> 10.1.1.3
ICMP TTL:64 TOS:0x0 ID:62868 IpLen:20 DgmLen:84
Type:0  Code:0  ID:19257  Seq:172  ECHO REPLY
=+=+=+=+=+=+=+=+=+=+=+=+=+=+=+=+=+=+=+=+=+=+=+=+=+=+=+=+=+=+=+=+

[]
```

- **$sudo snort -c /etc/snort/snort.conf**
- Watch the output carefully, and address any errors in your config file. (Hint: some default .rules files contain deprecated format, try to comment those files in the config file). Continue re-running snort until you get it working correctly.
- The command to run SNORT in IDS mode:

> - sudo snort -q -A console -c /etc/snort/snort.conf

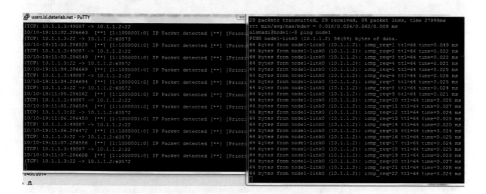

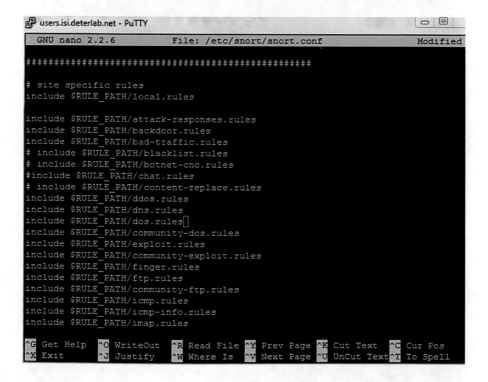

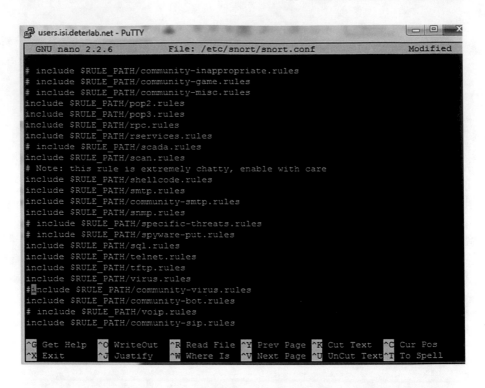

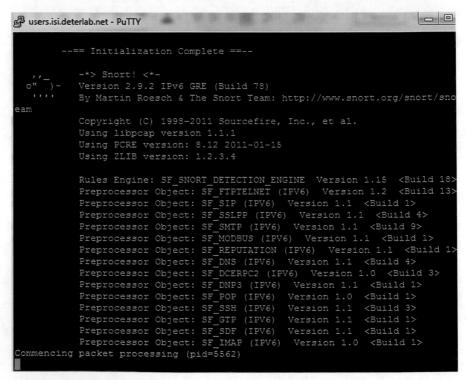

- Step 2: Read about Snort's signature syntax in the Snort User's Manual which is located on the class wiki. In particular, be sure to review the meta-data options reference and sid. Once you are somewhat familiar with the rule language, read through some of the web attacks rules files. These are files named in the form web-*.rules under **/etc/snort/rules/**. Follow the references listed in a few of the rules and read about the type of attack the specific signatures are designed to detect.
- The file (local.rule) is created for users to add their own rules (as its empty by default). We will add our experimental rules to this one
- Image below shows an example of 2 rules added to local.rules

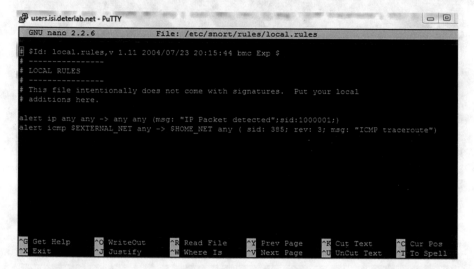

- Optional : You can create a new rule file (for testing)
- Try to add rules to your new rule file

• Make sure you add your rule file to config (you can do that in command line)

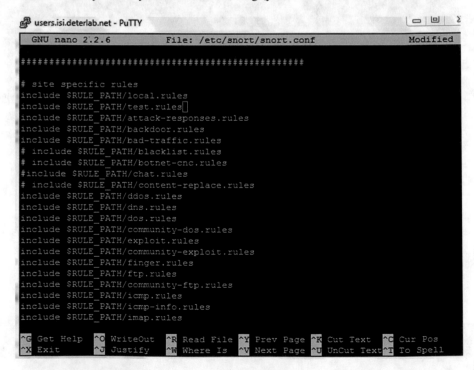

- Now, select two web attack signatures that seem straight-forward to understand. It would be simpler if you select a signature that looks for "evil" data in an HTTP URL string. Log into your Windows server and open a browser. Based on the documentation provided with the signature you have selected, attempt to trigger the Snort signature by making a HTTP request to which contains an attack string which should be detected.
- Next, verify in your Snort logs that your attack triggered an alert based on that. (Hint:/var/log/snort/)
- Step 3: Snort also allows us to write custom rules. Open the file /etc/snort/rules/ local.rules and add one rule that detects each visit to www.google.com that is made by the virtual machine. The rule should look for any outbound TCP traffic that is going to port 80 and contains the pattern "www.google.com" in the URL and trigger an alert when it gets a match. Give the rule an SID of 1000000 or higher. Then visit Google with a web browser and check if your rule triggered an alert. Record your screenshots as steps to write and test your (google) created rule

Lab related questions
1. In step 1, how did you modify the config file to make it work?
2. In step 2, describe the two attack signatures you chose and explain the corresponding rules against them. How did you attempt to trigger the alert? How did snort process your requests?
3. In step 3, copy/paste your new rule here. How did you confirm that your rule was enforced by snort?

Questions
- What is the difference between a switch and a hub? How can you practically check if this is a switch or a hub?
- Describe two examples of challenges in network forensics
- Can IP addresses be faked or changed? Can we detect such changes?
- How can we start a network forensic investigation?
- How could the information/log of a DHCP server be relevant to a forensic investigation?
- What is the difference between physical wire tapping and virtual wire tapping?
- Show with one practical example the problem with WEP encryption.
- What is the main difference between WEP and WAP?
- Is it possible to spy on a WiFi that is not broadcasting?
- How can "promisc mode" be useful for network forensics? How can it negatively impact network forensics?
- What is the different between a switch and a hub in terms of "promisc mode"?
- What are the different types in interfaces Wireshark can be used to listen to?
- Using Wireshark, how can you discover in the packets are using encryption or SSL?
- Using Wireshark, how can you discover a DNS query?
- Using Wireshark, how can you discover a DoS attack?
- Using Wireshark, how can you discover DHCP request?

- Using Wireshark, how can you extract packets based on certain size?
- How can "Follow a TCP stream" be useful for a network analysis?
- How can using "name resolution" help in network forensics?
- How can using "end-points" help in network forensics?
- How can using "statistics" help in network forensics?
- How can using "conversation" help in network forensics?
- How can you distinguish WiFi traffic using Wireshark?
- How can you focus analysis of a certain WiFi SSID using Wireshark filters?
- How can we use "Foremost" tool for analyzing PCAP files and what unique features it can be used for in network forensics?
- Which is easier to conduct network forensics with: a switch or a hub, and why ?
- What are the main artifacts that may include evidences in a switch?
- How can we detect "ARP spoofing" using Wireshark?
- How can we detect "port scanning" using Wireshark?
- How can "port mirroring" be helpful to network forensics?
- What are the main artifacts that may include evidences in a router?
- What are the main artifacts that may include evidences in a firewall?
- What are the main artifacts that may include evidences in an IDS/IPS?
- What are the main two forensic functions in which IDS/IPS can be used for?
- An IDS/IPS can distinguish malicious traffic using different techniques. Explain those different techniques.
- What filter can you use in Wireshark to extract unencrypted packets?

References

http://searchsecurity.techtarget.com/definition/network-forensics
https://www.sans.org/course/advanced-network-forensics-analysis
http://www.netresec.com/
https://securityintelligence.com/what-are-the-best-network-forensics-and-data-capture-tools/
https://www.amazon.com/Network-Forensics-Tracking-Hackers-Cyberspace/dp/0132564718
http://forensicswiki.org/wiki/Network_forensics
https://www.techopedia.com/definition/16122/network-forensics
https://www.corero.com/products/network-forensics.html
http://forensicscontest.com/
https://en.wikipedia.org/wiki/Network_forensics
https://www.savvius.com/elements/whitepapers/Network_Forensics_Security.pdf
http://www.networkcomputing.com/applications/network-forensics-separate-signal-noise/971953154
https://github.com/markofu/pcaps
https://wiki.wireshark.org/SampleCaptures

Chapter 12
Web Forensics-Chapter Competencies

Competency: Understand the investigation process for email communications
Activities/Indicators
• Study reading material provided by instructor related to major aspects of **investigating web activities.**
• Complete successfully an assessment provided by instructor related to competency content.
• For mastering levels, more than 80% of assessment grades should be earned in no more than three trials.
• Assessment questions can be pulled from the end of the chapter questions or any relevant material.
Competency: Understand the fraud techniques with email
Activities/Indicators
• Study reading material provided by instructor related to the types of fraud techniques associated with email communications.
• Complete successfully an assignment provided by instructor related to competency content.
Competency: Discuss the criminal and civil cases associated with electronic communications or web activities
Activities/Indicators
• Study reading material provided by instructor related criminal cases associated with electronic communications and web activities.
• Complete successfully an assignment provided by instructor related to competency content.
Competency: Discuss types of extraction and analysis
Activities/Indicators
Study reading material provided by instructor related to types of extraction and analysis with web forensics.
Complete successfully an assignment provided by instructor related to competency content.
Competency: Understand the process of evidence handling of emails and browsers information
Activities/Indicators
• Study reading material provided by instructor related to the process of handling digital evidence.
• Complete successfully an assignment provided by instructor related to competency content.

© Springer International Publishing AG 2018   283
I. Alsmadi et al., *Practical Information Security*,
https://doi.org/10.1007/978-3-319-72119-4_12

Overview

This chapter covers web forensic procedures commonly used by digital forensics professionals. The focus of the web forensics is on email communications, web browsing, and archived transmissions. The chapter explores the process of analyzing emails and web browsers when fraud or unauthorized activities have been identified.

Knowledge Section

Web Forensics

Organizations have to communicate with clients, suppliers, vendors, and other businesses. This communication can be a website, face-to-face, email, or by phone. In this chapter, we review investigations associated with emails and web browsing. As we have found, the process of forensics involves the identifying, collecting, reviewing, storing, and reporting the user activity. Therefore, it is important for the organization's two control and provide procedures for emails both internally and externally. This provides clarity to the employees on proper use and how digital transactions will be recorded if there is a violation. As with any procedure, we have basic steps to follow when an incident occurs. The following shows a brief outline of standards steps that can be followed during a forensic investigation.

Web Forensics Steps

1. Identification of the incident
2. Collection of evidence
3. Review evidence
4. Store evidence
5. Report results
6. Take necessary actions

The key is to have formal processes and clear procedures for consistency and possible litigation. There are times where this documentation may be used in a criminal case depending on how severe the activity is. A formal process will ensure each employee and instance is treated equally when there is an investigation. It is also a good practice to frequently review the overall procedures to ensure they are in line with current practices and external regulations.

Email

Email has been an important avenue for organizations to communicate. Many organizations have policies and procedures established for the employees. However, many times the practice of email is used inappropriately and not consistent with the organizations policies. In these cases, organizations have to investigate the activities associated with the employee or groups of employees. We have seen many instances where employees were terminated due to messages sent through emails or chat applications. It becomes more problematic when activities containing sensitive information is sent from the organization to another entity.

We will find mail client applications like Microsoft Outlook that are used for communications, but also experience many instances of SPAM and viruses. Therefore, it is important to have policies that help prevent external attacks along with educating the employees on proper use and expectations of the organization. The list below identifies email client options for both personal and organizational use.

Email Clients

- Microsoft Outlook
- Eudora
- Opera Mail
- Windows Mail
- Lotus Notes
- Send Mail
- Gmail
- Hotmail
- Yahoo
- Rackspace
- Zoho Mail

Additionally, organizations may host their emails on a mail server. The examples below provide a short list of available options in regards to the mail servers.

Mail Servers

- Microsoft Exchange Server
- hMail Server
- Icewarp Server
- Communigate Server
- Zimbra Server

Email Protocols

Emails are extremely efficient for communicating to clients, vendors, and suppliers. We can include text, audio, images, and videos with the transmissions. Protocols associated with email communications are Simple Mail Transfer Protocol (SMTP), Post Office Protocol 3 (POP3), and Internet Message Access Protocol (IMAP). Table 12.1 displays the protocols, descriptions, and syntax used with each.

The following Figures show the sections where IT personnel can include POP, IMAP, and SMTP settings. Figures 12.1 and 12.2 provides a display of a MS Outlook POP and IMAP account settings. This is usually set up with by IT, but knowing this configuration as an investigator can be extremely useful.

Table 12.1 Email protocols

Protocols	Description	Syntax
SMTP	Reliable and efficient way to send messages. Used by transfer agent to deliver messages.	HELLO, MAIL FROM, DATA, EXPN, HELP, NOOP, QUIT
POP3	Standardize approach to access mail boxes and download messages to local computer or server.	USER, PASS, STAT, LIST, RETR, DELE, QUIT, NOOP, RESET
IMAP	Provides the ability to access emails to server. Subset of POP to access emails from mail clients.	NOOP, STARTTLS, AUTHENICATE, LOGIN, LOGOUT, SELECT, EXAMINE, CREATE, DELETE, RENAME, SUBSCRIBE, UNSUBSCRIBE, LIST, LSUBM, STATUS, CHECK, CLOSE, STORE, FETCH, COPY, UID, CAPABILITY

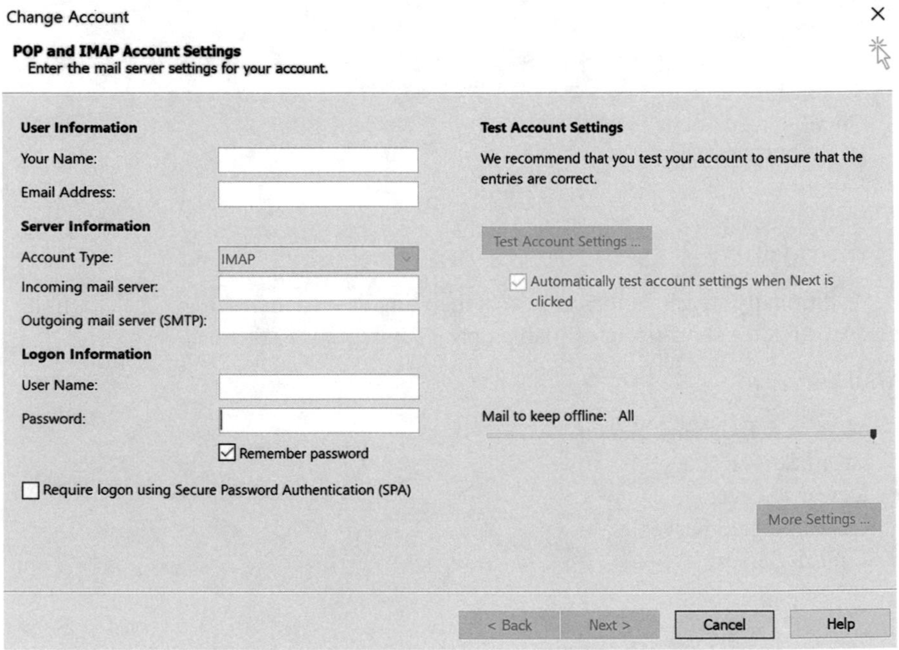

Fig. 12.1 POP and IMAP account setting with outlook

Email Forensics Tools

Once we have policies and procedures in place, then we move on to the collection of the activities. There are many tools we can use for the email forensics. However, some have a yearly licensing fee while others may be open source. It is best to review and examine each option before adopting the system for the forensic activities.

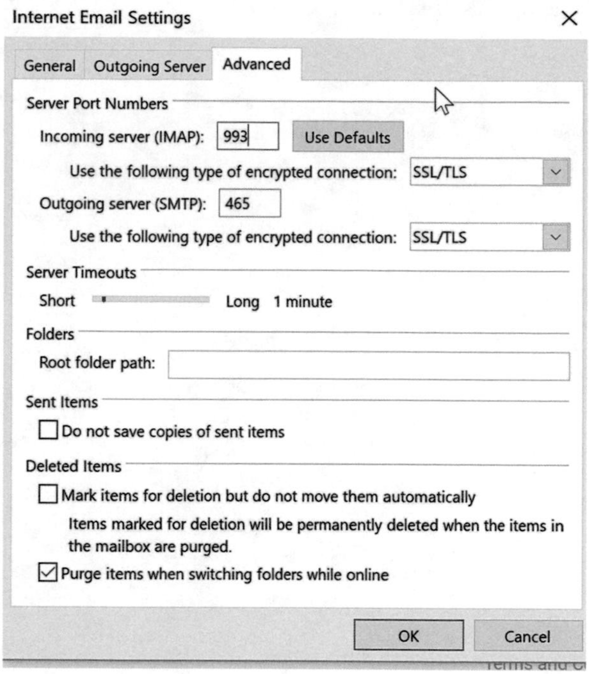

Fig. 12.2 Internet email settings with outlook

Table 12.2 Email forensic tools

Tools	Description	URL
MailXamier	Search for uncovered data to help respond to violations.	https://www.mailxaminer.com/
Paraben E3 Digital Evidence Examination	Powerful email examiner for forensics.	https://www.paraben.com/products/e3-p2c
Aid4Mail	Email forensics for email clients with a variety of formats.	http://www.aid4mail.com/
eMailTrackerPro	Trace email headers.	http://www.emailtrackerpro.com/

Table 12.2 provides a short list of potential tools Forensic Investigators can use during their analysis.

Email Headers

A forensic investigator may want to get access to the user's email account and then review the identifying information in the email header. The email will include information such as the header, content, and possible attachments. Table 12.3 shows the data we can obtain from an email header. Additionally, Fig. 12.3 provides a screen print of the Internet Header retrieved from an Microsoft Outlook email message.

Table 12.3 Outlook header fields

Received	Content-type
From	X-Mailer
To	X-MimeOLE
CC	Thread-Index
Subject	Return-Path
Date	Message-ID
MIME-Version	X-OrginalArrivalTime

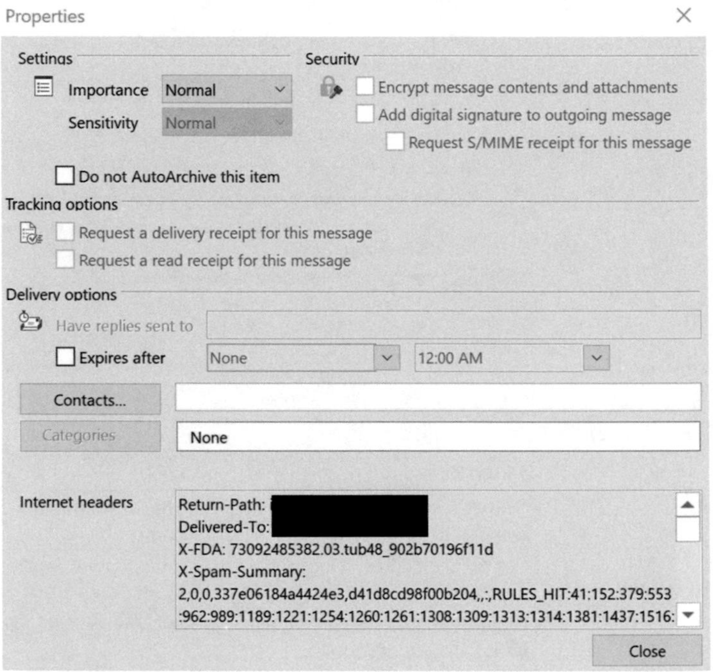

Fig. 12.3 Microsoft outlook message

Outlook Email Client

Yahoo Email Client

As we know, there are many email clients that organizations can use. Obviously, yahoo is not common for business operations, but is good to review the header content for comparison. The following provides the basic steps to retrieve the Yahoo header while Fig. 12.4 shows an example of a Yahoo header.

```
X-Apparently-To: burdwell@yahoo.com; Fri, 28 Jul 2017 16:37:13 +0000
Return-Path: <A/kfpjIXbTKCeKutz7PNkWA==_1126946551093_sxKSoN2lEeaJ9dSuUpzeEw==@in.constantcontact.com>
Received-SPF: pass (domain of in.constantcontact.com designates 208.75.123.225 as permitted sender)
X-YMailISG: vI4r7sMWLDtq4JC..JARuGTcqDZnMsaXEw7qnYccktwcL3fA
 2DB6TctQbssyvQFawReMz1zksvbnMyAoJjYhovhN.dIe_QIunUBTM7Wxi.AI
 rKH.SFHxMCqQXGleuG1squlHswHExaAIfJad8vHhqFhvBBY_bWG1kawux30h
 UqqLQj.8yUtyDYjqb6T1_z4ch5.6wuJsIiYSTD8gHwerSdWlQjQlcNC3Lpfl
 vRtM49K.e0ZigrWRE38mHb0EA_nTVGX3OO_H2rTZ9BnIdKRQ7E8T.6XvLIA.
 OnvHpivy386v8kPOcvFDbveDpyTq5jBv.SKOyFf8hYRoMUh6Qhk6.kn3pcWT
 yhTPM7MRz.bJC.WtEseOs.udzpXPw2s_Ur5VEPAqjGJrtEd3AoWVQbgxWgjo
 km4FlRky7_WQ8nfZiwbt4x_IPxFXQmTAdEFklaEPKDY.eslakKBYG3bMnWO7
 z1spo3pJef1670dWBJ8K49zIizwbiCkZKAVdbLTlYYkXNwdRXxpviVyiYREba
 4wmcVOGl4o.KG05yXYMGsd6eXhzWcOq993blAEUBGCscwGZxiSbeniZosUDU
 xnFJiLN6lllch4FlAf7yUdTCpd3Q6zv44v3MgMWISxp2nttbMCAn7l.wvotF
 HrLaUm8xz_04Dom9qFereelSpzXtpSXhhzfmtgic0HCh1EQoYmDv9MEzkHQW
 pr6obatgxfoCV_2MwxwMT15zUosJla5tBRc.1gubT.tfIXULPD599zcc92oF
 aEznsmVa4hBIn_oCJvmbPtyruKuqI5Pk0cH_3fgBjpLkKpGw9ClTTXcA7z3U
 gwZzSTYE5TL7lPSyUJLiZmJvbSPT4oMobjMuMtFwL0kWtrMA506AYke9tzF1
 wADOzvlnrilXR.TdND3GEk8C9NiAqGFpBiVhJOX6FGiQhCqtLPGE6JYfUJhR
 wSVuFzDhAiKV4yNKekFh1pu7QdUNpdqfEv.a0q8Nhd7Ny_wOrbMbOfUyrn0K
 .3ZIHHRoXfXtLTMNP_Qfq6q7rUVK_pv5nFx6YQDwv0WOyIeCS1FJ.qLuVFAR
 SyARwB8CEtqRAwZzfTzQ9Kadxcm2JQE16A7Xhb.gcQL_haD.TjOjN0lJ0QoN
 CSpa4.TqWHsPuoXhaPEk1hHpxq0Zto5uwdpyqCyURWf2lVQ.XbZ2JZbMdTsD
 aJJyX_FLwsYMg4q9moKFxxAw65Lh1nom3OaHIlpCh0frleJLXyqlOVxPcPr4
 nzSJMzG24NaNMXr36w62imIJ5bG4E.mia.jJUTyKV.EplaCHAW.Ap4ZjNnuY
 wRNg9PAvYTzXYpGb0rQkmlgzjI1h8WoSC_9h5ekogbYBjTnnn.YSISuKmPyr
 SPdxgQd.TlE70AgOUkH_353RvpL5TmTfzi_Q8e57pS_9m1kTUfMhL2OhR0Co
 fm8wLkDrTUN9ZgpX9lSoUl7zFWBhItiPJ5w8j1FWCcLac5sUoGNQqG32Xnpc
```

Fig. 12.4 Yahoo header

Yahoo Header Steps

1. Open the email
2. Click on … More
3. View Raw Message

Email Files

Many email clients save the email messages on a server or local drive. The following are a list of potential email formats that a forensics investigator may have to review. It is recommended to have a good working knowledge how to access these different email file formats within the organization. However, it may require for the investigator to also understand the other formats in case a potential employee or unauthorized individual is using different email clients for illegal activities.

Email File Formats
- PST
- OST
- EDB
- MSG
- MDS
- NSF
- EML
- DBX
- IMM

- MAILDIR
- ZDB
- IMAP
- MBX
- EMLX

As mentioned, we can utilize the features with the vendor email scanners to help retrieve source files associated with Outlook, EDB, Eduora, and Groupwise. Figure 12.5 shows an example of Paraben's feature to locate and examine the email database files. Please note the email database feature will differ depending on the email scanner vendor.

Once you complete the scan with an email scanner, then you will usually generate a summary table with folders and information on the activity. Table 12.4 shows a sample table of what an investigator will see as a summary of the email data file.

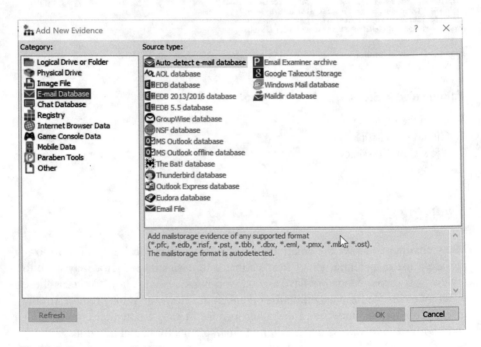

Fig. 12.5   Paraben e-mail database source types

Table 12.4   Email summary table

Folder name	Total messages	Folders	Unread	Size (bytes)	Attachments
Inbox	235	55	15	95,2541	13
Outbox					
Sent	55	55		43,254	15
Drafts	4				

Email File Reports

During the investigation, the forensic investigator will generate reports will all emails with content such as following. This information needs to be documented and organized to assist with the investigation.

- FROM
- Subject
- Body
- Attachments
- Recipients
- Date

It is key to note, there may be hundreds of emails generated with the report so it will take some time to review emails that require additional analysis. If the investigation is associate with multiple employees, then it is recommended to have several forensic investigators to make the correlations.

Web Browsers

Organizations use many web browsers such as Google Chrome, Firefox, Safari, and Internet Explorer. A Digital Forensic Investigator can investigation browser artifacts for internal employees and outside hackers. It is great to have an array of browsers to utilize, but this causes many issues for the investigator. For example, an investigator needs to understand where the Internet history and cache files are stored for each browser. To complicate the issue, the browsers are constantly updated so it is important for the investigating team to be up-to-date with the existing and new versions.

If we research top browser usage, then we will see Chrome is commonly used over Internet Explorer, Firefox, Edge, and Safari. However, we have to understand the usage of Internet Explorer, Firefox, Microsoft Edge, and Safari, since they account for the remaining half of Internet browser usage. When a forensic investigator reviews the browsers, then he/she can review the websites, timestamps, history, downloads, and cache files.

In Fig. 12.6, we see how Paraben allows the forensic investigator to review the cache data for Internet Explorer, Firefox, and Chrome. This is extremely useful when applications like this can assist with the investigation instead of manual reviews.

Criminal Cases with Web

We usually see many instances where there are several cases associated with hackers or employees accessing networks and sensitive information. However, we rarely see instances where organizations publicly announce employee had been investigated

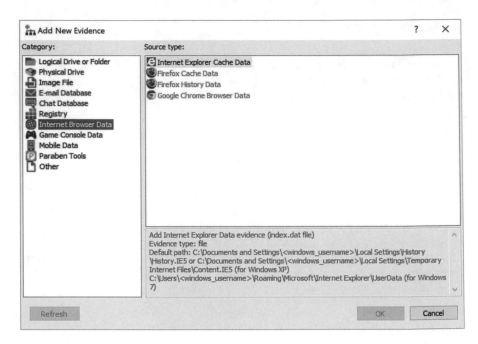

Fig. 12.6 Paraben internet browser data

within the organization. Forensic investigations should be familiar with the Electronic Communications Privacy Act, CAN-SPAM Act, Communication Assistance, to Law Enforcement Act (CALEA), Foreign Intelligence Act (FISA), and USA Patriot Act. Equally important is the organizations willingness and responsibility to work with outside law enforcement depending on the degree of the activity. At this time, the organization legal department will need to be notified to ensure adequate and proper steps are taken.

Skill Activities

Understand the Process of Evidence Handling of Emails and Browsers Information

Activities/Indicators

Scenario: Joe Bob has been suspected of sending sensitive information to a competitor. This incident was reported by another employee in the Sales department.

- Define the process of Collection, Examination, Analysis, and Reporting for an email incident.
- As the manager of the Sales Department, what actions should be taken to address this concern.

- Does the manager contact the IT department for further investigation?
- Outline what access will be needed to review such a case.
- Ensure to use the Author, YYYY citations with any outside content.

Resources
http://csrc.nist.gov/publications/nistpubs/800-45-version2/SP800-45v2.pdf

Review Email Headers

Activities/Indicators
Scenario: As a Forensic Investigator, you have been tasks to review the header of one Microsoft Outlook message that was sent from an employee in your organization to an outside entity.

- Open your email client.
- Create a new message called Sales Data.
- Send the message to an external email address.
- Review the email message header.
- Explain each section.
- Discuss how the section details can help with a Forensics investigation.
- Explain why it is important to communicate proper used of emails to employees.
- Ensure to cite appropriately using the Author, YYYY citations.

Resources
Outlook
https://support.office.com/en-us/article/View-e-mail-message-headers-cd039382-dc6e-4264-ac74-c048563d212c#bm2

Yahoo
https://help.yahoo.com/kb/SLN22026.html

Scan for Archived Files from Mail Client

Activities/Indicators
Scenario: You have been asked to retrieve and collect an email data folder with email files for an investigation.

- Download an open source or trial email scanner
- Open an email data file on the machine
- Perform the forensic analysis on the email message
- Explain what you discover during the investigation.
- What were the limitations with the tool?
- Why would an organization need to invest in a proprietary email scanner?

- Ensure to cite appropriately using the Author, YYYY citations.

Resources
https://digital-forensics.sans.org/community/downloads
https://www.mailscanner.info/
https://www.paraben.com/products/e3-emx

Review and Examine the Web Browsing History

Activities/Indicators
Scenario: As a Forensics Investigator, you need to review the browsing history of an employee in the Sales Department. The Sales Manager believes the employee is visiting unapproved sites on his work laptop.

- In MS Word, explain how you would handle the machine before the investigation.
- What steps would you take before logging into the machine?
- On your machine, download a Forensic Tool Kit
- Open the Web Browser
- Click on a few web pages to store the pages in the browser history
- Run the Forensic Tool Kit
- In MS Word, provide at least three screen prints showing the results. In the document, discuss what you found and how can this be used as digital forensics investigator.
- Ensure to cite appropriately using the Author, YYYY citations.

Resources
https://www.mcafee.com/us/downloads/free-tools/pasco.aspx
http://www.nirsoft.net/web_browser_tools.html
https://www.foxtonforensics.com/browser-history-viewer/
http://www.mountimage.com/
http://accessdata.com/product-download

Locate Essential Artifacts for Investigation (IE, Firefox, Chrome, and Safari)

Scenario: Create a document that explains the process of locating files, history, and activity on a browser.

- In MS Word, develop a document for the organization that explains the process of investigation browser activity for the organization.
- Identify the type of browser used in organizations.

- Ensure to cite appropriately using the Author, YYYY citations.

Discuss the Criminal and Civil Cases Associated with Electronic Communications or Web Activities

Activities/Indicators
- Research an article that deals with a criminal case associated with web activity and electronic communications.
- Provide a summary of the criminal case to include the activities and prosecution.
- Ensure to cite appropriately using the Author, YYYY citations.

Application Activities

Review and Examine the Web Browsing History

Activities/Indicators
- Download a Forensic Tool Kit.
- Open the Web Browser.
- Click on a few web pages to store the pages in the browser history.
- Run the Forensic Tool Kit.
- In MS Word, provide at least three screen prints showing the results. In the document, discuss what you found and how can this be used by a digital forensics investigator.

Resources
https://www.mcafee.com/us/downloads/free-tools/pasco.aspx
http://www.nirsoft.net/web_browser_tools.html
https://www.foxtonforensics.com/browser-history-viewer/
http://www.mountimage.com/
http://accessdata.com/product-download

Utilize a Web Forensics Software

Scenario: You are responsible for reviewing and analyzing the best forensic tool for your organization. Review forensic tools to see which tools provide the best options for emails and web browsers.

- The following are only options. You are not limited to only these vendors.

- FTK Imager
- Paraben
- AccessData's Forensic Toolkit

Resources

http://www.accessdata.com/product-download
https://www.paraben.com/
http://accessdata.com/products-services/forensic-toolkit-ftk

Questions
- Select the email forensic techniques.
- Select the web browsing techniques.
- Explain the process of extraction.
- Explain the sections or an email used for discovery.
- Select the processes for handling email and browser evidence.
- What file format is used with MS Outlook email files?
- Explain the process of collecting information for an email.
- What is included in the history log.
- Explain the process of examining the browsing history for IE and Chrome.

Chapter 13
Mobile Forensics

Competency: Know the different operating systems for the mobile devices

Activities/Indicators
- Study reading material provided by instructor related to major aspects of **identifying the different operating systems for mobile devices**.
- Complete successfully an assessment provided by instructor related to competency content.
- For mastering levels, more than 80% of assessment grades should be earned in no more than three trials.
- Assessment questions can be pulled from the end of the chapter questions or any relevant material.

Understand the information and data that can be examined on the mobile devices

Activities/Indicators
- Study reading material provided by instructor related to the data that can be examined during the digital forensics investigation of a mobile device.
- Complete successfully an assignment provided by instructor related to competency content.

Competency: Understand how the SIM card can be used for the investigation

Activities/Indicators
- Study reading material provided by instructor related to the investigation and analysis of the SIM card.
- Complete successfully an assignment provided by instructor related to competency content.

Competency: Understand how GPS coordinates can be used in an investigation

Activities/Indicators
- Study reading material provided by instructor related the GPS coordinates of a mobile device.
- Complete successfully an assignment provided by instructor related to competency content.

Overview

This chapter covers the forensic process with mobile devices. The forensics will explore the call history, mobile browser, and types of devices. The exercises will focus on the operating systems, device states, SIM, GPS, and device seize. It is important to cover these practices given the devices used by employees to conduct business functions.

© Springer International Publishing AG 2018
I. Alsmadi et al., *Practical Information Security*,
https://doi.org/10.1007/978-3-319-72119-4_13

Knowledge Section

Mobile Forensics

The use of mobile phones has dramatically increased over the past 10 years. Mobile usage is a common activity for both our personal and professional lives. Therefore, the acceptance and use of these devices in the work environment opens more possibilities for unauthorized access and unintended activities. These devices are extremely integrated into our daily work activities so there is usually enormous amounts of traffic located and stored on them. We can see an assortment of devices and operating systems ranging from Android to iOS along with the different devices. These devices provide Wi-Fi, battery power, touch screens, and app downloads. The devices can range from smartphones to tablets so organizations with many different configurations connecting to the network and business systems.

We see organizations adopting BYOD or issued mobile devices since it can help improve employee productivity. For example, the mobile device can help improve communications and data access to employees who work remotely in a geographical area. This becomes an efficient process for operations, but also opens up many possibilities for unintended use and criminal activities.

Data and transactions have been a target given the employees access to the networks and documents using the mobile devices. There are times when organizations need to recover digital evidence associated with the device contacts, photos, SMS messages. This information can be extremely valuable with digital forensics. For example, the phone history can be used to make an association between two parties collaborating for an unauthorized event. The following provides a list of what an investigator can view during his/her analysis.

Mobile Device Forensics
1. Call Logs
2. Contacts
3. Wi-Fi and GPS coordinates
4. SMS Text and MMS Video/Audio
5. Images
6. Email for email client
7. Folders

Organizations focus on the potential threats related to Wi-Fi, GPS, Bluetooth, SIM, and Near Field Communication (NFC). This needs to be a proactive approach with resources dedicated to protecting the infrastructure and network therefore reducing the amount of vulnerability opportunities. The following shows common features associated with smart phones and mobile devices.

Mobile Device Features
- Long Term Evolution (LTE) is used for data and voice communications. This allows for high speed access and transmission with 4G.

Table 13.1 Forensic investigator activities

Gather	Smart Phones	Cell Phones	Tablet and PDAs
Analyze	GPS Navigation Units	GPS Receivers	GPS Trackers
	Vehicle Navigation		

- Text messages is used with Short Messaging Service (SMS).
- Wi-Fi is based on IEEE 802.11. This is associated with the implementation of the wireless local area network (WLAN).
- Near Field Communication (NFC) use radio frequency for short range communications.

The mobile technology stack will include aspects such as hardware, firmware, application, and device. These categories will include the device memory, SIM, drivers, data, and type of device.

As we know, organizations protect against these potential issues, but there will be times where the forensics investigator will need to review each of these areas. Let's examine potential activities a forensics investigator may have to perform. It is key to also know that some of these activities may be performed at the request of the organization or as a request by a law enforcement agency. Table 13.1 Displays potential activities that can be performed from a digital investigation perspective.

Mobile Device Hardware and Software

Mobile devices have RAM, ROM, hardware, and interfaces. We can usually transition from one device to another given some of the similarities. These devices have the ability to install mail clients, VPN connections, and cloud applications. There is an array of hardware associated with the process, memory, display, text input, camera, GPS, wireless, and battery. Although, these are not always investigated, it is good to know the configurations in case further analysis is needed. The software of mobile devices includes the operating systems, Personal Information Manager (PIM), apps, text messaging, chat, email, and web browser.

Operating Systems

There are plenty of options when it comes to the types of operating systems associated with the mobile devices. There is Android, iOS, Windows, and Blackberry along with other less used environments. This poses a problem given the digital forensics examiner has to be familiar with the particular operating system for the mobile device. There are variances and an array of differences between the

Table 13.2 Mobile device operating systems

Device	Operating system
Android Samsung	Android OS Jelly Bean, Ice Cream, Gingerbread, Honeycomb
Apple iPhones	iOS
Nokia, Samsung Focus, and HTC Titan	Windows

operating systems. Additionally, there are differences with the hardware, touch-screen connectivity, messaging, folders, storage, and applications. Therefore, a forensics investigator will need to have a working knowledge of accessing the different operating systems and layers associated with the mobile device used in the organization. Table 13.2 list the type of devices and operating systems commonly found in organizations.

As can be seen, there are many variations of devices along with continuous updates and editions with the operating systems. As a result, it is critically important for the forensic investigators to consistently update their skills with each device.

Device States

There are times when the mobile device is in a Device Discover, Allow, Block, Quarantine, Upgrade, Nascent, Active, Quiescent, or Semi-Active state or mode. This will be important if a forensic investigator is conducting an ongoing investigation. For instance, the user may block his/her activity to the exchange, therefore, removing it from the network device detection. As a forensics investigator, it will be this individual's responsibility to understand the different states as the investigation is carried out.

SIM Card

The Subscriber Identity Module (SIM) is a circuit card that stores the identity and subscriber identity of the mobile device. The card can store contacts, personal information, and temporary information. A complication to this particular card is the ability for the user to exchange the SIM card. This can be problematic during a digital forensic investigation since the user may have exchanged the SIM card prior to the investigation. The SIM card will contain the main file, which is the RAM, ROM for temporary and permanent storage. The challenge is understanding if the files or activity is stored locally on the phone or in the SIM card.

Table 13.3 GPS file formats

Raymarine	SIMRad
GPS Exchange Formats (EPX)	Magellan
KML	Standard Horizon
JSON	NMEA Data Sentences
Garmin	TomTom

GPS

Global Positioning System (GPS) coordinates the mobile device position using satellites with the Global Navigation Satellite Systems (GNSS). There will be times when a forensic investigator will need to review track logs, way points, routes, stored locations, and recent addresses. This information can be critical in whether a potential employee is engaged with unauthorized activities depending on the situation. There are thousands of profiles associated with the file formats, memory, and images with the GPS. Additionally, forensic software vendors provide tools to import the file types from Garmin, Magellen, TomTom, Raymairne, and SIMRad. Again, this can be challenging because of the different types from these GPS providers. Table 13.3 provides a sample lists of GPS provider file formats.

Device Seize

The process of seeking a mobile device from an employee can be extremely risky. We have to remember this employee will be pointed out as conducting activities that were not in line with the organizations policies. Therefore, the organizational policies need to be in place and communicated to all employees so everyone is aware of the expectations. It is recommended to request access from the employee for the investigation opposed to just seizing the device. This will provide consent from the employee for the investigation. However, the decision to investigate the device will come with some confrontation at some point regardless if the employee is guilty or not. At this point, the trust will be broken between the employee and organization. Knowing this, the investigator needs to be certain that have been violations with the organizational policies.

Mobile Forensics Tools

Many investigators will utilize manual skills and knowledge to review the mobile devices during an investigation. However, it is critically important to also have tools and software available for the investigation. This will allow the investigator to uncover hidden transactions or data always viewable by an individual. NIST (2014) states the forensic tools need to be useful, comprehensive, accurate, deterministic, verifiable, and tested. Remember, these devices can contain years of data. The following shows a short list of forensic tools that can be used for the investigation. However, there are many open source options, but they need to be vetted before being used in an organizational setting.

Possible Forensic Tools
- Paraben
- SANS Investigative Forensic Toolkit (SIFT) Workstation
- Secure View
- Oxygen Forensics
- Data Doctor
- Sim Card Data Retrieval Utility
- Device Seizure F
- Forensics SIM Cloner

Figure 13.1 shows Paraben's options for acquiring and adding evidence for the investigation. Other vendor options will have similar interfaces to add the device and then collect the evidence.

Figure 13.2 shows Paraben's option for selecting the Mobile Data source for the investigation. As shown, there are options for acquisitions, imports, and back up files when examining the mobile data.

Once the analysis is conducted, the forensic investigator will see device information, install apps, device contacts, call log history, and other findings. Figures 13.3, 13.4, 13.5, and 13.6 show sample sections that can be generated by a forensic toolkit. In this approach Paraben E3 generated information such as timestamps, SID,

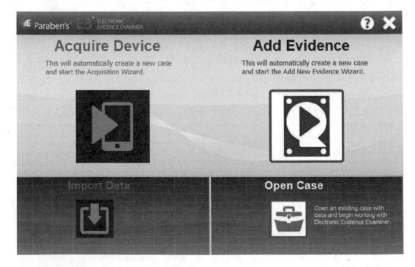

Fig. 13.1 Paraben acquire device and add evidence

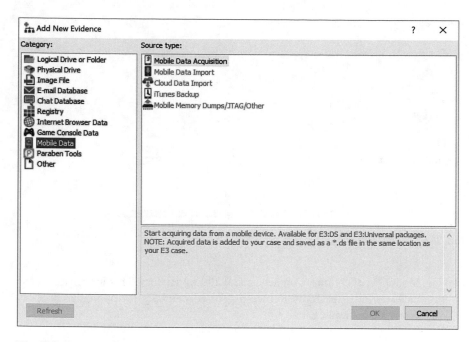

Fig. 13.2 Paraben mobile data

Program timestamp	10/13/2016 6:10:32 PM
Device ID (IMEI)	A000002CF1F1C8
Subscriber ID (IMSI)	000007742681764
Device Software Version	MSM8626BP_1032.355.66.00R
Line 1 Number	7742681967
Network Operator Name	Searching for Service (00000)
Network Type	Unknown
Phone Type	Unknown
SIM Operator Name	Boost Mobile (311870)
Board	MSM8226
Brand	motorola
Device	falcon_cdma
Firmware ID	KXB20.9-1.10-1.18-1.1
Model	XT1031

Fig. 13.3 Device information

Installed Applications

Icon	Application Name	Version	Category	Manufacturer	Malware Suspicious	Internal Application Name
	Chrome	53.0.2785.124	Communication	Google Inc.	Suspect	com.android.chrome
	Facebook	97.0.0.18.69	Social	Facebook	Suspect	com.facebook.katana
	Jott	1.2.0.0	Communication	Juxta Labs, Inc.	Low Suspect	com.jott.android.jottmessenger
	Kik	10.16.1.9927	Unknown	Kik Interactive	Low Suspect	kik.android
	Maps	9.38.1	Navigation	Google Inc.	Low Suspect	com.google.android.apps.maps
	Messenger	92.0.0.13.70	Communication	Facebook	Highly Suspect	com.facebook.orca
	Snapchat	9.41.0.0	Social	Snapchat Inc.	Suspect	com.snapchat.android

Fig. 13.4 Installed applications

Device Contacts

Fig. 13.5 Device contacts

Fig. 13.6 Call log history

Device, Model, Applications, Contacts, and Call Log History. This information can be extremely useful if the forensic investigator is searching for particular calls or activity on a specific application.

Skill Activities

Know the Different Operating Systems for the Mobile Devices

Activities/Indicators
- Define and Outline the different operating systems for mobile devices used in business operations.
- Explain the features associated with each device operating system.
- Discuss why an organization may use one device over another.
- Ensure to properly cite any outside content.

Identify the Layers of an Operating System for a Mobile Device

Activities/Indicators
Scenario: Your organization has employees that use both Android and iOs devices for business operations.

- In MS Word, identify the operating system layers of both Android and iOs.
- Diagram both as part of an investigation.
- Explain the components identified in the diagram.

- How does the investigation differ when examining the text messages from each operating system?
- Explain best practices for accessing each operating system.
- Ensure to properly cite any outside content.

Resources

http://nvlpubs.nist.gov/nistpubs/SpecialPublications/NIST.SP.800-124r1.pdf
http://csrc.nist.gov/publications/drafts/nistir-8144/nistir8144_draft.pdf

Outline the Steps for Examining the Device Folders on a Mobile Device

Activities/Indicators

Scenario: You have been asked by your supervisor to create a procedure for examining mobile devices with a focus on the device folders for files and images stored inside the mobile device folder.

- In MS Word, create a procedure to examine device folders.
- The procedure will need to focus on Android or iOS.
- Identify manual activities and software that can be used to complete the examination.
- Ensure the document acts as a report for the organization.
- Ensure to properly cite any outside content.

Resources

http://nvlpubs.nist.gov/nistpubs/SpecialPublications/NIST.SP.800-101r1.pdf
https://www.cftt.nist.gov/AAFS-MobileDeviceForensics.pdf

Identify the States of a Mobile Device

Activities/Indicators

Scenario: You notice there is no policy in place to address the mobile states during discovery or forensics in the department.

- In MS Word, create a document that identifies the possible mobile device states that exist in the organization.
- Explain how a forensic investigator can adjust the state of a mobile device.
- Ensure to properly cite any outside content.

Resources

http://nvlpubs.nist.gov/nistpubs/SpecialPublications/NIST.SP.800-124r1.pdf
http://nvlpubs.nist.gov/nistpubs/SpecialPublications/NIST.SP.800-101r1.pdf

Understand how the SIM Card Can Be Used
for the Investigation

Activities/Indicators
Scenario: An employee is under review for storing unauthorized images of the enterprise network. The intent of these image captures is unknown, but requires and investigation since there is no reason determined by management that this employee should have these images on his work mobile device.

- In MS Word, create the steps needed to review the SIM card.
- Outline the information that can be obtained from the SIM Card investigation.
- Ensure to properly cite any outside content.

Resources
https://nccoe.nist.gov/sites/default/files/library/mtc-nistir-8144-draft.pdf
http://resources.infosecinstitute.com/sim-card-forensics-introduction/#gref

Understand How GPS Coordinates Can Be Used
in an Investigation

Activities/Indicators
- In MS Word, identify how GPS coordinated and be used to criminate or exonerate an employee if there was an investigation regarding the location of an employee during a particular time.
- Ensure to properly cite any outside content.

Document How GPS Can Be Used in an Investigation

Activities/Indicators
Scenario: As a forensic investigator, you have been asked to develop procedures for tracking GPS locations if an employee is ever investigated. This is a proactive and planning measure for any future cases.

- Identify the tools and techniques used for such an investigation.
- Determine how the extraction of the GPS locations can be obtained
- Explore potential tools or consultant services to help with the investigation.
- Ensure to properly cite any outside content.

Resources
https://nccoe.nist.gov/sites/default/files/library/mtc-nistir-8144-draft.pdf

Create a Process for a Company to Seize a Mobile Device from an Employee

Activities/Indicators

Scenario: Your organization does not have standard procedures for seizing a device from an employee.

- In MS Word, develop a forensic policy for seizing a mobile device from the employee. At this point, the investigation will initiate so it is important to have clear and consistent steps in place to ensure no violations of discrimination. This cannot be an unreasonable search for the employee.
- The fourth amendment does not protect the employee from searches, but this must be done with careful and reasonable consideration.
- Ensure to properly cite any outside content.

Resources

https://www.littler.com/five-lessons-employers-california-v-riley

Application Activities

Examine the Device Information, Installed Apps, Call History, Contacts, and Messaging

Activities/Indicators

Scenario: You are a forensic investigator. Your department has seized a mobile device from a current employee. You will need to perform a forensic investigation on the device and report the findings.

- In MS Word, explain the initiating steps for your investigation.
- Download a Forensic Vendor application to conduct the digital forensics.
- Connect your mobile device for the Review the device information, installed apps, call history, contacts, and messages.
- Use the software report generate as evidence
- In MS Word, create a report indicating your findings along with a recommendation on the results.
- Ensure to properly cite any outside content.

Resources

http://nvlpubs.nist.gov/nistpubs/SpecialPublications/NIST.SP.800-101r1.pdf
https://www.cftt.nist.gov/CFTT-Booklet-08112015.pdf
https://www.magnetforensics.com/mobile-forensics/
https://www.paraben.com/product-categories/mobile-investigations
https://www.paraben.com/products/e3-ds

Open Source Options

https://www.basistech.com/autopsy/
https://www.nist.gov/sites/default/files/documents/forensics/6-Mahalik_OSMF.pdf

Questions
- Identify the different mobile device operating systems.
- What information can be reviewed on a mobile device. Explain the process of accessing this information?
- Discuss what can be reviewed on a SIM card?
- Discuss the potential liabilities associated with mobile device forensics.
- Outline the steps for gathering GPS data from a device.
- Explain how forensics differs with Android compared to iOS devices?
- How can an investigator identify a device on a wireless network?
- Outline the steps for unlocking a protected mobile device.

References

https://nccoe.nist.gov/sites/default/files/library/mtc-nistir-8144-draft.pdf
http://nvlpubs.nist.gov/nistpubs/SpecialPublications/NIST.SP.800-101r1.pdf

Erratum to: Practical Information Security

Izzat Alsmadi, Robert Burdwell, Ahmed Aleroud, Abdallah Wahbeh, Mahmood Al-Qudah, and Ahmad Al-Omari

Erratum to:
I. Alsmadi et al., *Practical Information Security*,
https://doi.org/10.1007/978-3-319-72119-4

The volume published with the coauthor Mahmood Al-Qudah's name that appeared as "Mahmoud Ali Al-Qudah" which is incorrect. The book has been revised with his correct name, Mahmood Al-Qudah.

The updated online version of this book can be found at
https://doi.org/10.1007/978-3-319-72119-4

© Springer International Publishing AG 2018
I. Alsmadi et al., *Practical Information Security*,
https://doi.org/10.1007/978-3-319-72119-4_14

Index

© Springer International Publishing AG 2018
I. Alsmadi et al., *Practical Information Security*,
https://doi.org/10.1007/978-3-319-72119-4